Engineering Swarms of Cyber-Physical Systems

Melanie Schranz
Lakeside Labs, Klagenfurt, Austria

Wilfried Elmenreich
University of Klagenfurt, Klagenfurt, Austria

Farshad Arvin
Durham University, Durham, UK

CRC Press
Taylor & Francis Group
Boca Raton London New York

CRC Press is an imprint of the
Taylor & Francis Group, an **informa** business
A SCIENCE PUBLISHERS BOOK

First edition published 2026
by CRC Press
2385 NW Executive Center Drive, Suite 320, Boca Raton FL 33431

and by CRC Press
4 Park Square, Milton Park, Abingdon, Oxon, OX14 4RN

Library of Congress Cataloging-in-Publication Data (applied for)

ISBN: 978-1-032-04715-7 (hbk)
ISBN: 978-1-032-04733-1 (pbk)
ISBN: 978-1-003-19446-0 (ebk)

DOI: 10.1201/9781003194460

Typeset in Times New Roman
by Prime Publishing Services

For Product Safety Concerns and Information please contact our
EU representative GPSR@taylorandfrancis.com Taylor & Francis
Verlag GmbH, Kaufingerstraße 24, 80331 München, Germany

Contents

To Kennedi, my forever reason.

BLOOMSBURY ACADEMIC
Bloomsbury Publishing Inc, 1359 Broadway, New York, NY 10018, USA
Bloomsbury Publishing Plc, 50 Bedford Square, London, WC1B 3DP, UK
Bloomsbury Publishing Ireland, 29 Earlsfort Terrace, Dublin 2, D02 AY28, Ireland

First published in the United States of America 2026

Cover design by Jason Enterline
Cover image © Beeldbewerking/iStock/Getty Images Plus

Library of Congress Cataloging-in-Publication Data
Names: Gadson, Danielle N. author.
Title: Two steps forward, three to the side : social image and the dance for community health center survival / Danielle N. Gadson.
Description: New York : Bloomsbury Academic, 2026. |
Includes bibliographical references and index.
Identifiers: LCCN 2025032904 (print) | LCCN 2025032905 (ebook) | ISBN 9781666926699 hardback | ISBN 9781666926712 paperback | ISBN 9781978762398 eBook | ISBN 9798216265658 epdf | ISBN 9798881891497
Subjects: LCSH: Community health services–United States.
Classification: LCC RA445 (print) | LCC RA445 (ebook)
LC record available at https://lccn.loc.gov/2025032904
LC ebook record available at https://lccn.loc.gov/2025032905

ISBN: HB: 978-1-6669-2669-9
ePDF: 979-8-216-26565-8
eBook: 978-1-9787-6239-8

Typeset by Deanta Global Publishing Services, Chennai, India

For product safety related questions contact productsafety@bloomsbury.com.

To find out more about our authors and books visit www.bloomsbury.com and sign up for our newsletters.

Two Steps Forward, Three to the Side

Social Image and the Dance for Community Health Center Survival

Danielle N. Gadson

BLOOMSBURY ACADEMIC
NEW YORK • LONDON • OXFORD • NEW DELHI • SYDNEY

FSC
www.fsc.org
MIX
Papier | Fördert gute Waldnutzung
FSC® C083411

Kolibri 360 GmbH
Ferdinand-Jühlke-Straße 7
99095 Erfurt, Deutschland
produktsicherheit@kolibri360.de

Preface

In the evolving landscape of modern technology, Cyber-Physical Systems (CPS) represent the forefront of innovation, blending the physical and digital worlds seamlessly. However, as these systems become increasingly interconnected and complex, new challenges arise. In complex systems composed of numerous interacting components, the necessity for adaptation or expansion often becomes evident. Minor adjustments, such as a simple software update, might subtly alter component behavior, undermining key assumptions about the broader system. How can we design systems that are not only powerful but also robust, adaptable, and scalable? Conventional top-down design approaches often fall short meeting these requirements, particularly when addressing the ongoing design, operation, and maintenance of such complex systems. One promising answer lies in the fascinating field of swarm intelligence.

Swarm systems utilize simple local interactions to produce sophisticated, emergent behaviors similar to those observed in natural entities like flocks of birds or colonies of ants. Inspired by this potential, *Engineering Swarms for Cyber-Physical Systems* aims to guide readers through the entire journey of applying swarm intelligence in CPS, from conceptual modeling to final deployment. This book is structured to be a one-stop resource for engineers, researchers, and students. It begins with an introduction to swarm intelligence, illustrating its vast potential through examples from robotics, manufacturing, and search-and-rescue operations. We then explore the core principles of modeling, design methods—including machine-learning approaches—and simulation techniques. A highlight of the book is its hands-on character: programming examples and practical insights are woven throughout, enabling you to translate theoretical concepts into real-world applications effortlessly.

Our aspiration is that this book will serve not only as a technical guide but also as a source of inspiration. Whether you are an experienced engineer or a curious researcher, you will find the tools and knowledge needed to create innovative swarm-based CPS solutions.

Thank you for taking this journey with us. Together, let's explore the future of engineering swarms and unlock the potential of intelligent, adaptive systems.

Happy designing!

Contents

List of Figures

List of Tables

List of Acronyms

ACO	Ant Colony Optimization
AES	Active Elastic Sheet
ANN	Artificial Neural Network
API	Application Programming Interface
BDD	Block Definition Diagram
CPS	Cyber-Physical System
CPSoS	Cyber-Physical System of Systems
CPU	Central Processing Unit
EA	Evolutionary Algorithm
FREVO	Framework for Evolutionary Design
FSM	Finite State Machine
GUI	Graphical User Interface
IBD	Internal Block Diagram
IC	Integrated Circuit
ID	Identifier
IoE	Internet of Everything
IoT	Internet of Things
IMU	Inertial Measurement Unit
IR	Infrared
ISP	In-System Programming
MCU	Microcontroller Unit
NEAT	NeuroEvolution of Augmenting Topologies
OS	Operating System
PCB	Printed Circuit Board
PSO	Particle Swarm Optimization
PCT	Planned Cycle Time
RFID	Radio-Frequency Identification
ROS	Robot Operating System
RPT	Raw Process Time
SAR	Search and Rescue
SCXML	State Chart XML

SoC	System on Chip
SPP	Self-Propelled Particles
SysML	Systems Modeling Language
TCACS	Tabu Continuous Ant Colony System
TMS	Traffic Management System
UAV	Unmanned Aerial Vehicle
UI	User Interface
UGV	Unmanned Ground Vehicle
UML	Unified Modeling Language
URDF	Unified Robotic Description Format
USV	Unmanned Surface Vehicle
UUV	Unmanned Underwater Vehicle
UWB	Ultra Wideband
WIP	Work in Progress
VTL	Velocity Template Language
V2X	Vehicle to Everything

Chapter 1

Swarm Intelligence in Cyber-Physical Systems

Swarm intelligence is a concept adapted from nature to complex technical systems of various application domains including smart grids, cities, or mobility, and industry 4.0. Swarms of bees, birds, fish, and other organisms demonstrate how a self-organizing behavior can solve complex tasks without a central control entity dictating rules and goals to the individual agents within the swarm. This chapter aims to give a general overview of the typical terminologies for swarm intelligence concepts in general and motivates the engineering of swarm intelligence for CPSs.

1.1 Basic Terminology

A swarm consists of individual, simple, and homogeneous agents [86]. Swarm behavior describes social animal behaviors that exhibit a strong innate or developed inclination to collaboratively achieving a common global objective, such as foraging, nest-building, or defending against enemies. Thus, swarm intelligence is based on the coordination and control mechanisms that exist in natural swarms, which operate in dynamic and diverse environments. Despite the complexity of their goals, individual agents in a swarm typically follow simple rules and interact locally with their peers and surroundings. By doing so, they generate collective behavior that enables the swarm to solve complex tasks. This makes the swarm scalable, adaptable, and robust to changing conditions.

In nature, swarm behavior is often observed in animals that exhibit eusocial behavior, like honeybees, ants, termites, and naked mole rats [275]. Eusociality is characterized by several distinctive features, including cooperative care of offspring, where individuals within the group collaborate

to nurture offspring, extending their care even to those that are not direct descendants. Additionally, eusocial species often exhibit a clear division of labor, with distinct castes or groups specializing in specific tasks within the colony. Another hallmark of eusociality is the reproductive division of labor, where reproduction is taken care of by a subset of the individuals within the group, often a reproductive queen and a few males. This specialization in reproduction ensures a well-organized and efficient social structure. Furthermore, a key characteristic is the presence of overlapping generations in the swarm. This stands in contrast to species where parents typically perish before the offspring reach maturity. These features collectively define and distinguish eusocial behavior in various animal species. Given proper abstraction, eusocial behavior can be an inspiring model for CPS swarm systems.

A very early definition of a swarm was rather simple, referring to a group of animals executing a joint movement pattern. This is still reflected in the traditional definition that can be found in the Oxford Dictionary, where a swarm is *"A large or dense group of flying insects"*. The first formal definition was given by Farley and Clark in 1954 as "a system which changes its basic structure as a function of its experience and environment" [110]. The term *swarm intelligence* was first introduced in 1989 by Beni and Wang [33]. It was used to describe the dynamics of a group of cellular robots that could execute an intelligent collective behavior. This marks the starting point at which swarm intelligent behaviors were studied outside of natural sciences [333]. Bonabeau and Meyer [40] in 2001 effectively summarized all the necessary characteristics of a natural system to call it a swarm intelligence system: *"Social insects work without supervision. Their teamwork is largely self-organized, and coordination arises from the different interactions among individuals in the colony. Although these interactions might be primitive (one ant merely following the trail left by another, for instance), taken together, they result in efficient solutions to difficult problems (such as finding the shortest route to a food source among a myriad of possible paths). The collective behavior that emerges from a group of social insects has been dubbed swarm intelligence."* The extension of the same concepts to swarm robotics was captured by the definitions of Şahin and Spears [320] in 2005: *"Swarm robotics is the study of how a swarm of relatively simple physically embodied agents can be constructed to collectively accomplish tasks that are beyond the capabilities of a single one"* and *"Swarm robotics emphasizes self-organization and emergence while considering the issues of scalability and robustness."*

Animals exhibiting various evolved swarm behaviors to collaboratively achieve common goals, such as foraging, nest building or defending against enemies have long inspired swarm intelligence in robotics and advancements in CPS developments. The observed swarm intelligence models gain an understanding of the principle patterns and rules that the agents execute in a natural system. This also gives insights into the conditions, rules, and interactions that lead to a swarm behavior. Moreover, research has shown that an individual member of a swarm is typically incapable of finding an

optimal solution on its own [131]. Instead, a successful solution often arises from the collective behavior of the swarm as a whole.

From a CPS perspective, an engineered swarm comprises a collection of agents that can vary in complexity from relatively simple entities like swarm robotic research platforms with small, basic robots equipped with few sensors to highly complex systems such as autonomous cars. As a swarm, these agents reach a common global goal collaboratively using relatively straightforward behavior rules, which are implemented locally by each agent. When individuals interact with each other and/or their environment following well-designed rules, they can exhibit collective behavior capable of solving complex tasks effectively. Systems designed based on swarm intelligence typically exhibit characteristics, such as parallel and distributed processing and control, scalable performance, adaptability to dynamic variations, and resilience to losses and failures of individual components [333].

1.1.1 Swarm-Intelligent Behavior

To determine whether a system is truly swarm-intelligent, we propose considering the following two cases [333].

1. If the system's functionality remains intact even when operated by a single swarm member it does, not qualify as a swarm-intelligent system. If there is only one active agent, the system retains its full set of capabilities. In this scenario, these capabilities do not depend on any swarm behavior, even if multiple agents are concurrently in operation [333]. An illustrative case is the concept of "sweeping," also known as uniform coverage, which can be achieved by a single swarm agent or multiple agents collaborating to cover a designated area. To address a coverage problem within a fixed area using a UAV swarm, a simple approach is to distribute the UAVs across the area. Each UAV independently executes a sweeping algorithm for its assigned section of the map, without any interaction with other swarm members during the algorithm's execution. The sole drawback of employing a single UAV in this context is that it might take longer to complete the task.

2. As the swarm size increases, the overall performance ratio improves, indicating the presence of a swarm-intelligent system. If the overall performance per member decreases as the swarm size increases across the entire size range, the system faces scalability issues rather than benefiting from larger population sizes. For instance, increasing the number of cars on a street boosts the capacity to transport people and goods. However, as cars interfere with each other, the incremental gain in transport capacity diminishes once a certain density is reached. Thus, in this scenario a car's operation does not benefit from the presence of additional vehicles.

A comprehensive examination of this test, along with supporting empirical data, can be found in the work by Hamann et al. [153]. In such a swarm-intelligent system, the agents reach their global objective according to locally executed rules from which the overall behavior emerges through the inter-swarm interactions. In the case of CPS swarms, the behaviors displayed by each individual CPS are based on a local set of rules. These rules can be as simple as mapping sensor inputs to actuator outputs or as complex as evaluating and analyzing local information before taking action. These behaviors require interactions with the physical world, which encompass the environment as well as other CPSs [122]. To perform these interactions, the CPS reads and interprets sensory data, processes it, and drives the actuators accordingly. This sequence of interactions is known as the basic behavior, which is repeated either indefinitely or until a desired state is achieved [333].

1.1.2 A Taxonomy on Biologically-Inspired Swarm Intelligence Algorithms

Swarm intelligence algorithms are inspired by the field of biology. Most algorithms mimic the behavior of a natural system and apply it to an engineering problem. Thus, swarm intelligence algorithms are usually referred to as nature-inspired [155]. Apart from biology, other research fields like physics and chemistry also inspire the development of algorithms. However, in the proposed taxonomy, these fields are not considered. Although these algorithms are nature-inspired they do not demonstrate swarm intelligence behavior. This also holds for evolutionary computation, self-organizing neural networks, and cellular automata. There are other frameworks that draw inspiration from nature or natural processes but are not considered swarm-intelligent. Among them are biomimetic algorithms, which rely on mimicking biological processes and models. Examples include flower pollination algorithm [416], great salmon run [259], and dolphin echolocation [196]. Other more generically nature-inspired algorithms mimic physical or chemical laws instead of biological ones. Some examples of these types of algorithms include simulated annealing [206], spiral optimization [375], water cycle algorithm [107], and galaxy-based algorithm [349]. Other notable examples include the fireworks algorithm and its variants [376], the swarm chemistry particle algorithms [323], and various variants of the harmony search algorithm [129]. These examples exhibit the characteristic properties of swarm intelligence, but their local rule set does not have a specific biological source of inspiration. Given the vast array of inspiration sources and the ocassionally indistinct boundaries between algorithms, it can be challenging to pinpoint categories or processes that can be definitely identified as recognized sources of inspiration for swarm-intelligent behavior [333].

In this section, we exclude the theoretical and mathematical descriptions of individual swarm intelligence algorithms, as these have already been extensively covered by various authors including Bonabeau et al. [39],

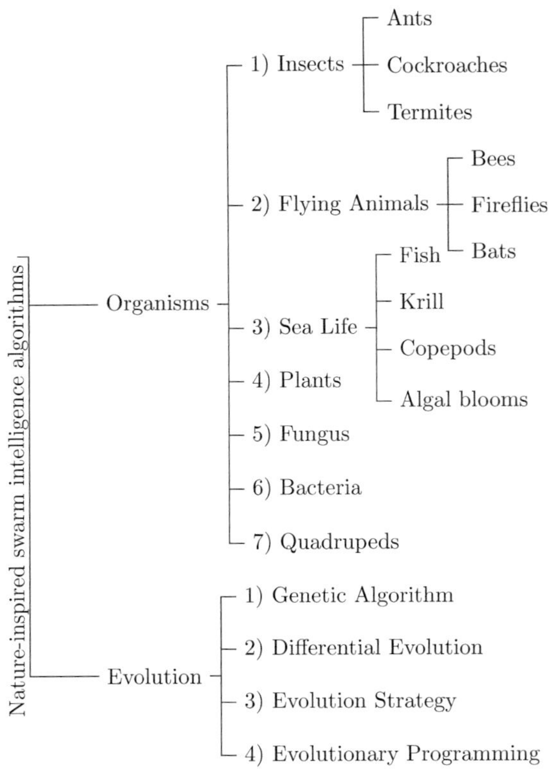

Figure 1.1: Taxonomy of nature-inspired swarm intelligence algorithms.

Camazine et al. [53], Garnier et al. [131], Blum and Li [36], Floreano and Mattiussi [122], Parpinelli and Lopes [281], Binitha and Sathya [35], Yang et al. [417], Krause et al. [214], Hassanien and Emary [155], Yang et al. [418], among others.

For classifying swarm intelligence algorithms, this is not accurate enough as many nature-inspired algorithms actually do not relate to swarm behavior [145]. Therefore, we propose the taxonomy of models for existing and possible future swarm intelligence algorithms based on their sources of inspiration, as seen in Figure 1.1. The purpose of this categorization–organisms and evolution–is not to present a complete taxonomy of swarm intelligence inspirations. Instead, it aims to highlight the diversity of natural inspirations and emphasize that only a few swarm intelligence algorithms are currently suitable for use in swarms of CPSs. The biggest issue is their applicability for swarms of CPSs. For instance, Ant Colony Optimization [88] would require individual CPSs to utilize the environment for information exchange through stigmergy. Possibly future use of swarming CPSs might turn this requirement into an advantage.

Swarm intelligence is not limited to natural systems alone but can be applied to any complex system comprising multiple interacting components that can exhibit useful properties like collective decision-making, regulation, homeostasis, and periodic patterns. Thus, the key to utilizing any collective system-whether composed of living components or not-as a source of inspiration for designing a swarm-intelligent system lies in understanding why and how it functions effectively [333]. Thus, behaviors as described in Section 1.1.1 should be extracted, abstracted, and translated to the CPS swarm domain.

1.1.3 Multi-Agent vs. Swarm Systems

Both terms appear frequently in literature, making it challenging to distinguish between multi-agent and swarm systems. Characteristics typical of one system often appear in the other. A common feature is that both multi-agent systems and swarm systems consist of multiple agents or CPSs that cooperate. Additionally, attributes such as interaction types, local intelligence, and distributed knowledge are partially assigned to both systems.

Therefore, categorizing systems into a single class is often impractical. In the following sections, we will further analyze characteristics that provide sufficient evidence to classify a system as multi-agent, swarm, or, as is often the case, irrelevant in the first place.

Emergence and Homogeneity

Swarm systems are considered to be bio-inspired drawing their behavior from organisms like ants, bees and fish, etc. In such systems, we argue the swarm members are numerous, relatively simple, and homogeneous [333]. The simplicity of each agent implies that a single swarm member would be incapable of achieving the overall system's goal independently. Additionally, the global goal is typically unknown to the individual agents. This is where the concept of emergence becomes crucial: The interactions and local rules with limited local knowledge of the simple swarm members lead to an emergent global behavior through collective interaction [333].

In multi-agent systems, it is assumed that each agent can accomplish the task independently. It (i) simply takes longer if we only use one instead of many agents, or (ii) only a portion of the global goal will be achieved due to the agent's heterogeneity or specialization. Thus, the individual agent has the ability to reach the global goal without necessarily relying on emergent behavior.

Distributed Operation

Another key characteristic of swarm systems is the lack of a central control that leads to emergent behavior. Due to the simple and local rules and the

interactions among the agents, a swarm system is, per se, a distributed system with distributed knowledge [333].

For a multi-agent system, distributed processing is not mandatory. The agents can also be controlled centrally such that parts of the task are fulfilled in a distributed manner.

1.2 Motivation to Engineer Swarm Intelligence for CPSs

With the increasing number of interconnected CPSs, individual and hierarchical control becomes impossible. The latest Study of Juniper Research in 2018 suggests that the quantity of interconnected devices will increase steadily and reach 50 billion by the year 2022[1]. In contrast to static environments, the growing interconnectivity of components in modern and future technical systems has increased the complexity of system design and operation in dynamic environments. This necessitates greater adaptability, flexibility, and robustness. The systems that integrate multiple interacting components exist at various scales, ranging from System on Chip (SoC) that form the computational core of ubiquitous modern devices like smartphones, to Internet of Things (IoT), which links billions of edge devices, and the Internet of Everything (IoE), which encompasses people, processes, data, and devices. In essence, the world is transforming into a truly interconnected and collective realm, characterized by the proliferation of systems that combine numerous interacting components at different scales [333].

In this respect, CPSs have emerged as a leading domain for exploring and implementing multi-component systems, characterized by a strong interconnection between computational (software-side) and physical (hardware-side) resources [224]. The National Science Foundation defines CPSs as "Engineered systems that are built from, and depend upon, the seamless integration of computation and physical components.". They integrate various domains, such as sensing, computation, control, and networking, into physical objects and their infrastructure, enabling interaction [123]. Thus, CPSs represent a major paradigm in the framework of collective and connected systems, as well as a vertical study of systems: They are inherently transdisciplinary, generalizing and expanding individual sub-fields such as embedded systems, robotics, and networking, by simultaneously merging concepts from cybernetics, mechatronics, design, and process science [226, 333]. Additionally, aspects related to decision autonomy, system integration, cyber-security, control, scalability, optimization, validation, and verification, play a major role in CPSs' design and control. Furthermore, CPSs function within the physical world, which is characterized by constant dynamic changes, unpredictable events, and external conditions that are difficult to

[1]Study of Juniper Research, `https://www.juniperresearch.com/press/press-releases/iot-connections-to-grow-140-tohit-50-billion`, June 2018.

model, and the involvement of other CPSs and human agents [333]. Hence, the CPS model is well-suited to describe and reason on the plethora of complex and interacting components that are being deployed, connected, and integrated with our everyday lives. This results in a multitude of interconnected and interacting components that are best characterized by the concept of Cyber-Physical System of Systems (CPSoS), which refers to a vast and distributed complex system of CPSs. In essence, this creates an ecosystem of CPSs that operate at multiple scales and interconnect with one another [274]. In a CPSoS[2] model, the design and control challenges become even more complex as individual component autonomy must be integrated with explicit considerations for interdependence and coordination, interoperability, distributed control, and emergence of behaviors. This high complexity necessitates methods that should achieve the following goals:

- Distributed control, supervision and management.
- Local coordination among the composing subsystems.
- Partial autonomy of the subsystems.
- Capability of dynamic reconfiguration of the system as a whole on different time scales.
- Evolution of the overall system during its operation.
- Possibility of generating useful emerging behaviors at the system level [104].

From the above-mentioned list of goals and characteristics for CPSs, it is evident that they align with the concept of a swarm system and the typical features expected in a swarm-intelligent designed system. Applying swarm intelligence to CPSs is not an entirely new idea. Significant progress has been made in the swarm robotics domain in this regard, particularly in recent years [153, 332]. However, applying the concepts of swarm intelligence to more general CPSs, particularly to real-world CPSs that are large, heterogeneous, multi-scale, and autonomous, remains a challenge yet to be fully mastered.

For the successful design and development of a swarm intelligence solution for CPS swarms, it is important to consider both the physical and cyber aspects of the involved CPSs. The optimization, networking, and physical embodiment aspects of these systems require the seamless integration of software agents, mechatronic devices, and communications, unifying them into a cohesive whole [224, 226]. A CPS can be modeled as a swarm consisting of multiple components, with each component

- integrating one or more physical devices (sensors, actuators, communication, memory, processors, etc.),

[2]Given that a CPSoS is a CPS itself, in the following, we will use the acronym CPS to refer in a general sense to either a single CPS or to a CPS swarm to make a distinction, if relevant.

- acting autonomously (i.e., control is distributed and/or decentralized),
- responding and possibly adapting to changing conditions, and
- locally communicating and interacting with other swarm components to possibly produce effective and useful behaviors at the system level [333].

Nevertheless, two fundamental problems can be identified in the design and deployment of CPS swarms [333]:

1. Formalization of CPSs from a swarm intelligence perspective: This process involves extracting and generalizing properties and parameters that are shared among various CPSs, which are significant for designing swarm-intelligent behavior.
2. Open research topics of the swarm intelligence domain itself (see Chapter 7 for more details).

A good example of a CPS is the consideration of an autonomous car as done in Schranz et al. [333]: At a smaller scale, each of the major components of the car (e.g., the anti-blocking system) can be described as CPSs. At a larger scale, a fleet of autonomous cars interacts and communicate with each other. This can be used as a representation of a CPS swarm. To obtain real-time sensor and traffic data, the cars in the swarm will need to communicate with other CPSs that are present in the surrounding environment and the supportive infrastructure. These CPSs can constitute an additional swarm that facilitates interaction between different swarms of CPSs. Since there is no specific limit to the number of devices or swarms, it is possible to create a large-scale, multi-level swarm system that incorporates hierarchies where required (e.g., to a central monitoring entity), integrating many CPSs that interact with multiple CPS swarms.

Chapter 2

Principles of Swarm Intelligence

A swarm system belongs to the family of self-organizing systems. Swarm intelligence is derived from the natural behavior exhibited by social animals. Their behavior tends to be adaptive, robust, and scalable [52]. These are desirable properties of an autonomous system that can be replicated with swarm intelligence approaches in the design of real-world technical systems. All three properties add additional behavioral patterns to systems that use swarm intelligence (see Figure 2.1). Additionally, a swarm is characterized by a critical number of members, micro rules and macro plans, positive and negative feedback, interactions through direct or indirect communication leading to emergent behavior, super linear effects and the ability to balance exploration and exploitation.

In summary, a group or collection of individuals can be classified as a swarm when it displays swarm behavior that includes all the aforementioned properties and characteristics.

2.1 Micro and Macro Level

In a swarm-intelligent system, local (micro) level changes drive the global (macro) level behavior through bottom-up causation. This allows agents to function in a coordinated manner without the need for central control.

There are two main approaches for modeling a self-organizing system, following the micro-level (also known as local or microscopic) and the macro-level (also known as global or macroscopic) perspective [164]. On the micro level, each system agent has its own local state space and local behavior. Hence, each agent is explicitly represented. Macro-level modeling does not look inside each agent but has a global perspective. Thus, the macro level

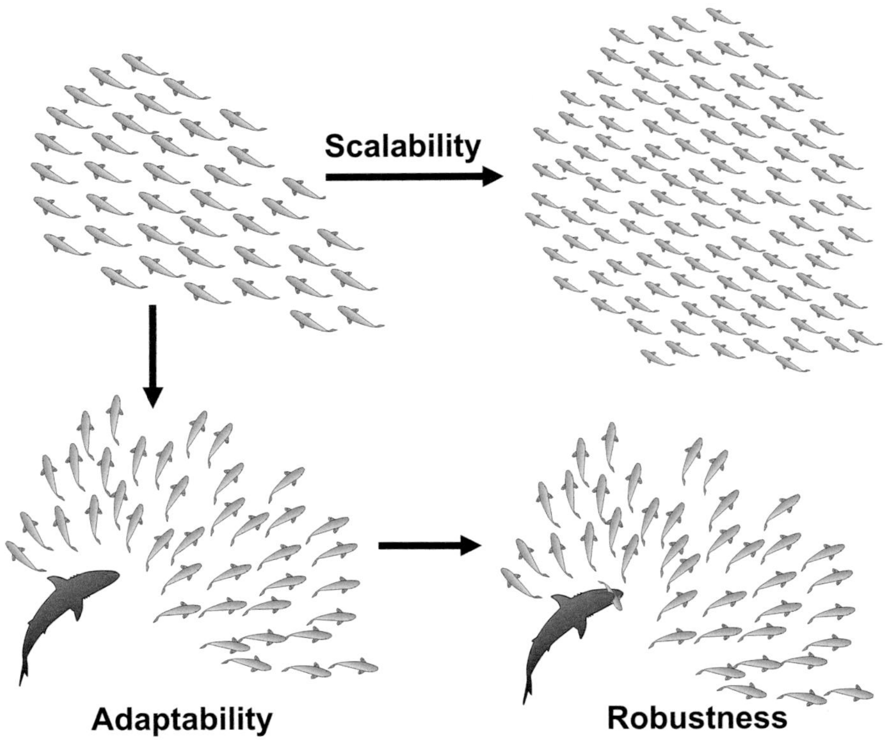

Figure 2.1: The advantages of a swarm: Adaptivity, robustness, and scalability (adapted from [292]).

only considers the global state of interest and how this state changes over time. Detailed knowledge of an agent's operation is thus not required for the macro level. It is more important to model the level of abstraction to work on a global task level. Thus, the macro level abstracts agents' details away, e.g., their position or the state in the sequence of action.

Especially for modeling swarms of CPSs, the distinction between the micro and macro levels plays a crucial but contrasting role. The overall goal is defined at the macro level, involving the operation of multiple swarm agents, while each agent follows behavior defined at the micro level.

2.2 Emergence

In the design of a technical system the main question arises with regard to predictability of the system's behavior. Additionally, predictability is related to controllability that affects the ability to control the system and achieve a desired configuration and/or output. Due to the property of emergence, predictability and controllability are properties that are inherently hard to achieve in swarm-intelligent systems that follow a bottom-up approach. Thus,

emergence is directly correlated with micro and macro behaviors. Emergence is generated with local decisions and local interactions on a micro level that lead to a system function on the macro level. The most important characteristic is that this emergence can produce completely new properties that were unimaginable when considering only the micro or the macro level [153]. This aligns with Aristotle's principle that "the whole is greater than the sum of its parts". Thus, understanding the indivdiual components of a self-organizing system is only part of the solution. While these components can be simple and easy to comprehend it is their interplay that leads to emergence, creating entirely new solutions [32].

In swarm systems, predictability and controllability are not the only challenges; any parameter or configuration setting can significantly impact the resulting dynamics of a swarm-intelligent system.

2.3 Feedback

Interactions between agents are characterized by either synergy (cooperation or positive sum) or friction (conflict or negative sum) [161, 140, 38], also known as feedback loops in complex systems theory [273]. These feedback loops lead to non-linear dynamics of swarm systems, thereby creating complex behavior.

Agents are assumed to be goal-directed: They try to maximize their fitness, satisfaction or utility. When both agents gain fitness, the interaction is said to be synergetic. When both lose fitness, the interaction is characterized by friction. When one gains while the other loses, the interaction is competitive. If the losses match the gains, it is considered zero-sum. Since agents seek to maximize their fitness, they prefer synergetic interactions, and try to avoid friction. However, since interactions are local, agents do not a priori know which interactions with other agents will be most beneficial for the overall behavior. They can discover this through trial-and-error, which mirrors the evolutionary dynamics of blind variation and natural selection of the fittest (most synergetic) interactions.

Nevertheless, friction is not always a negative characteristic. Positive feedback typically increases fitness, or other parameters that could lead to permanent growth [153]. Still, at a certain point of time increasing values do not produce any growth, but rather start disturbing each other as the resources are finite. This is where negative feedback has its positive role that stabilizes the systems final state. This again could lead to local maxima or minima that necessitate monitoring a self-organizing system's behavior.

2.4 Scalability

Swarm intelligence algorithms are able to produce complex and scalable swarm behaviors from simple and local rules, thus, on a micro-level. Scalability is a crucial aspect of a swarm system, enabling it to perform effectively with

varying numbers of swarm members and problem sizes. As noted in [85], a swarm is able to maintain its function and interaction among its parts even as its size increases, without the need for a redefinition of its interaction mechanism. This means that the addition or removal of swarm members will not result in a significant decline in performance, as long as the number of members does not fall below a certain critical mass[1] [333]. One cornerstone of implementing scalability is to restrict the agents' communication to the interaction with their local neighborhood instead of to all agents. In contrast, an all-to-all broadcasting mechanism will likely break scalability due to communication demands increasing superlinearly with swarm size.

2.5 Adaptivity

Adaptivity represents the ability of a swarm to adapt to dynamic environments, to cope with different tasks that could also appear unforeseen, and to still fulfill its mission and reach the macro-level goal. Thus, an adaptive system uses a sequence of operators applied through an adaptive plan to produce a performance that is tailored to and observable within the environment. These sequences produce different system responses dependent on the environment. Thus, the selection of the right, applicable sequence of operators is the task of the adaptive plan that defines the performance, i.e., the fitness from the operations on the environment [163].

2.6 Robustness

The vision of autonomous systems is to design a robust system behavior that can cope with a variety of unforeseeable errors and perturbations without human supervision. The reason for this is a paradigm shift from monolithic systems to large networked systems with many independent and constantly changing components. Robustness is the ability of a swarm to adapt to variations in swarm size, environmental changes or other disturbances affecting the system or the environment. Redundancy is a mechanism through which biological systems adapt to their environments [371]. In technical systems, redundancy involves cost considerations and, for static systems, management challenges.

The term robustness has different meanings dependent on the community in which it is used [221]. In the swarm robotics community, robustness is the ability of a system to cope with erroneous input[2], system errors, disturbances or attacks affecting specific agents or subsystems during execution [114]. This leads to other advantages, for instance, the dependability of a system.

[1] It should be noted that it is not yet clarified what this critical threshold (the minimum number of swarm members) should be [153].

[2] 1990. IEEE Standard Glossary of Software Engineering Terminology, IEEE Std 610.12-1990 defines robustness as "The degree to which a system or component can function correctly in the presence of invalid inputs or stressful environmental conditions."

Generally, in a distributed system errors and small variations in input do not affect the correctness of the system's execution process. Moreover, the system as a whole exhibits graceful degradation: When arbitrary agents or links are removed or damaged, the quality of the output is likely to deteriorate, but very gradually. Thus, damaging, subverting or removing an agent may still leave the overall function intact. This is what produces robustness. This is not the case in a centralized system's control, where any malfunction of the controller represents a possible single point of failure.

Up to now, robustness is more or less a "descriptive" characteristic, without formally defined measures. As we can see in the following contributions by different authors, there is no formally defined technique that ensures the robustness of a swarm with respect to either global or local behavior.

2.7 Communication

Autonomous agents in a swarm need to continuously exchange information. Even a single byte, such as an agent's velocity or relative position, is essential for maintaining autonomy in a swarm system.

Communication in a swarm system can occur both directly and indirectly. In the direct communication an agent uses a communication medium such as Wi-Fi or infrared to transmit information to its immediate neighbor. This can be a one to one message to a single neighbor or a one to many message to several surrounding neighbors. Indirect communication involves an agent generating a message and sharing it with the environment, for example, through an access point, allowing other agents to collect the message either randomly or intentionally. The shared communication medium might be a specific location where agents deposit and retrieve messages. Indirect communication methods like depositing pheromone values, are commonly used by social insects and other animals for survival, whereas direct communication through message exchange is prevalent in the robotics domain.

Pheromone-based communication is one of the most prevalent indirect communication methods used by social animals [411]. Pheromones are complex chemical substances that facilitate intraspecific communication by being deposited into the environment by an agent and detected by others. These chemicals convey intricate messages that trigger behavioral and developmental changes within a swarm. This form of communication is widely adopted by various animals ranging from yeast and small insects like ants and bees to mammals such as dogs and humans [120, 294]. In swarm systems pheromone communication has been simulated using substances like alcohol [125], phosphorescent paint [240], light [133, 18], and infrared [285, 391]. Figure 2.2 illustrates a bio-inspired artificial pheromone-based system that mimics ants pheromone-based communication.

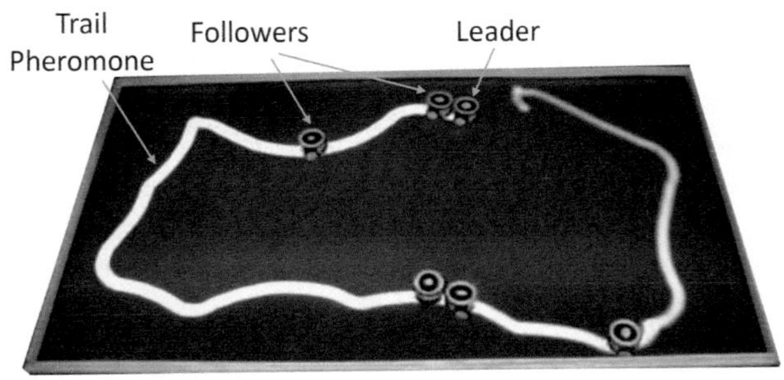

Figure 2.2: Example of an artificial pheromone system, COSΦ [18], that was developed for swarm robotic applications. In this scenario, a leader is releasing pheromones and five followers are following the leader. The pheromone trail is evaporating over time.

2.8 Superlinearity

We anticipate enhanced performance in a swarm due to synergy effects and other factors including:

- **Synergy effects:** When agents collaborate they can achieve more than the sum of their individual capabilities. Their interactions can lead to innovative solutions that isolated agents could not achieve.
- **Task specialization:** Agents in a swarm can specialize in specific tasks performing them more efficiently allowing the swarm to benefit from diverse expertise and skills.
- **Parallel processing:** Tasks can be distributed among agents, enabling them to work simultaneously. This can significantly reduce the time needed to complete complex tasks.
- **Adaptive behavior:** Swarms can adapt to changing environments or tasks dynamically, optimizing their performance over time based on feedback and learning.

If the performance of a single agent is given as an execution time T_1 to complete the task, we would expect multiple agents to perform the task faster. In the following, we define performance *Pf* as the reciprocal value of the execution time of a task:

$$Pf = \frac{1}{T}$$

In a perfectly scaling architecture, we would expect that n agents are able to fulfill a task in $T_n = \frac{T_1}{n}$ time.

There are cases where the collaboration of agents provides a synergistic effect. In such cases, the performance even exceeds linear scaling, so that

$T_n < \frac{T_1}{n}$. An example of such a task is a scenario in which a swarm efficiently solves a complex puzzle through collective problem-solving. Each agent contributes to discovering different parts of the solution, significantly speeding up the process. A more dynamic, though admittedly ferocious, example is a hunting party tracking elusive prey. A single agent, or in this case, a hunter, would be very inefficient since the prey could easily elude the hunter. However, a coordinated group of hunters would be significantly more efficient.

In an extreme case, the performance of a single agent could be even zero while being a positive number for the swarm. Consider a scenario where robots need to clean an area by pushing obstacles out of the way. If a single robot encounters an obstacle that is too heavy to move on its own, it won't be able to complete its task. However, with the cooperation of multiple robots these obstacles can be successfully pushed aside.

Such cases are considered to exhibit *superlinear performance*. Increasing the number of agents in a group is anticipated to result in the group accomplishing more work compared to a smaller group under the same conditions and timeframe. In a group exhibiting real swarm-intelligent behavior, such a superlinear characteristic is highly desired. This characteristic indicates that the effect of the overall system is greater than the sum of the effects of its individual parts. The effective design of swarm-intelligent behavior relies on the synergies of cooperation among individual swarm members. Only when these interactions enhance the performance of each member can the overall system be considered well-designed. This means that within the bounds of a feasible swarm size, not only the efficiency of the whole swarm but also the efficiency of each individual swarm member has to increase. We refer to this as the *swarm effect*,

$$\frac{Pf_{\text{large\_swarm}}}{\text{size}_{\text{large\_swarm}}} > \frac{Pf_{\text{small\_swarm}}}{\text{size}_{\text{small\_swarm}}}, \tag{2.1}$$

where Pf_i represents any quantitative performance metric of the swarm i in which larger metric values identify a higher or better swarm performance and size_i represents the size of that swarm, given by the number of swarm members. As indicated in the formula, the average performance per agent increases with the swarm size provided the swarm is performing effectively. This principle is shown in Figure 2.3 that illustrates the expected performance scaling properties of three different systems:

(a) The performance of an imaginary algorithm or swarm model that scales linearly with $O(n)$ as the dashed line in Figure 2.3, where n is the number of swarm members, without considering physical constraints or limitations, is not practical in the real world.

(b) The performance scaling of an algorithm that accounts for sharing resources with other algorithm instances or a swarm model that views

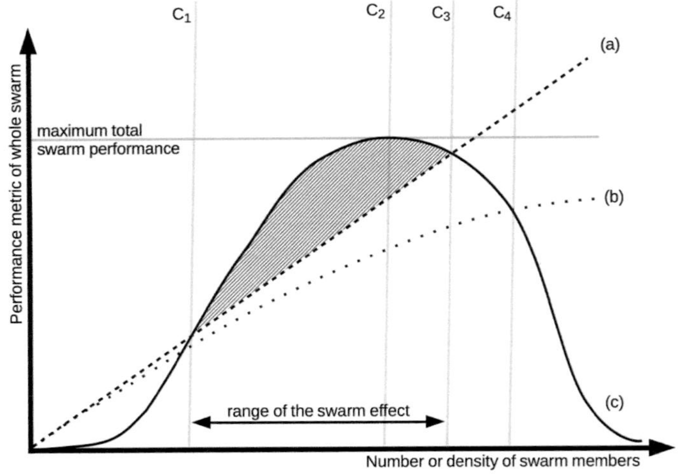

Figure 2.3: The swarm performance illustrated as a function of the number of swarm members, extracted from various experiments [37, 153]. Line (a) shows the hypothetically linear scale of a disembodied algorithm with swarm size. Line (b) presents the logarithmic scaling of a disembodied algorithm that requires shared resources among its instances when implemented for a parallel execution. Finally, line (c) shows the typical scaling of physically embodied swarms (or of physics-considering simulation of such agents) [333].

space as a shared resource for agents will scale with $O(\log n)$ as the dotted line in Figure 2.3, or even slower, due to the increased communication and coordination overhead that comes with a larger number of members.

(c) The solid hat-shaped curve depicted in Figure 2.3 represents the performance of a physically embodied swarm system operating in the real world.

With a very low swarm size or density of agents $N < C_1$, the swarm shows almost no performance. This can be traced to the low connectivity between the agents that is not high enough to allow for cooperation and interaction. As the size of the swarm increases, the performance of the system improves due to the swarm effect, where interactions between members lead to greater efficiency. Once the swarm reaches a size of $N > C_1$, it can take advantage of these interactions to scale in a super-linear fashion. This high level of performance can be maintained within a certain size range of $C_1 > N \geq C_3$, and even beyond C_3, the performance remains above a logarithmic scale. However, the overall peak performance of the swarm system is reached at a threshold of C_2, beyond which the addition of more swarm members results in a decline in average performance due to crowding and other factors. Once the swarm

size surpasses another critical level of $N > C_4$, the performance of the system deteriorates and eventually leads to a total loss of performance.

Adding agents to perform a task together typically adds overhead due to several factors:

- **Additional communication:** As the number of agents increases, the amount of information that needs to be exchanged among them also increases. The demand for richer communication for the swarm to function effectively [243] can lead to higher latency and potential bottlenecks if the communication infrastructure is not optimized for scalability.
- **Coordination of movements:** Ensuring that all agents move and act in a coordinated fashion requires complex algorithms that can introduce delays. Misalignment in timing or direction can reduce the overall effectiveness of the swarm. Offline optimization of tasks [5] is limited due to uncertainties in environmental parameters. For example, a robotic drone could be delayed by wind conditions or unexpected obstacles.
- **Resource contention:** Multiple agents may need access to shared resources, such as data or physical space, which can lead to conflicts or inefficiencies if not managed properly using assignment algorithms [271].
- **Increased complexity in decision-making:** With more agents, the complexity of decision-making increases, as there are more possible interactions and dependencies to consider [245]. This can slow down response time and reduce overall efficiency.

Note, however, that a swarm is not always formed for performance reasons. For example, it could be acceptable for a swarm system to have a dropping performance for an increasing number of agents if other qualities are prioritized, such as:

- **Robustness and dependability:** A swarm system is less likely to fail completely if individual agents are capable of taking over the tasks of failed members, ensuring continued operation.
- **Maintainability:** With a modular design, individual components or agents can be easily exchanged or repaired without disrupting the entire system.
- **Extensibility:** If there is future demand, it is possible to enhance the system's performance by adding more agents, allowing for scalable solutions to problems as they arise.

2.9 Exploration and Exploitation

Swarm members being loosely coupled, can operate in two distinct modes. In *exploration* mode they disperse to explore the environment or, in case

of an optimization algorithm [276], to explore a solution space to gather information. Conversely, in *exploitation* mode, they utilize the gathered knowledge to maximize reward, such as acquiring food or identifying the optimal solution within the explored area.

One typically needs to strike a balance between exploration and exploitation to achieve optimal results. If the exploration part is emphasized too much, swarm members can get lost in a large search space without providing a valuable contribution. Conversely, a swarm that concentrates excessively on exploitation might overlook more lucrative areas. Exploration and exploitation do not have to follow a strict sequence; some algorithms like Ant Colony Optimization [90] include both concepts simultaneously. The problem of exploration and exploitation is also not limited to swarms, but is a common problem whenever there is uncertainty in a system that needs to be explored at a cost. A famous example for this conflict in probability theory and machine learning is given by the multi-armed bandit problem [195], where numerous slot machines, each creating rewards according to a specific, but unknown probability distribution, need to be explored and then exploited for maximizing gain.

Chapter 3

Modeling Swarms of Cyber-Physical Systems

A model of a system is a simplified representation derived from an actual (technical) system. Modeling a swarm of CPSs helps to lower the level of abstraction, streamline and formalize the technical system. We aim to identify a representation for each agent that is familiar to us and meets the microlevel requirements necessary for contributing to macro-level swarm behavior [248]. During abstraction we omit parts of the CPS to get a simplified system that is easier to understand. This is a typical task for engineers who must discern while still capturing the overall CPS mission. To enhance this understanding, we formally define the mission using logical descriptions and mathematics [153]. Concluding with a formal description of the system, we aim to understand how the self-organizing system behaves under various configurations and potentially in different environments.

In a self-organizing system γ, where $\gamma \in \Gamma$ with Γ as the configuration space, we can only observe the actual configuration at a specific, discrete time step t with $\gamma_t, \gamma_{t+1}, \gamma_{t+2}, \ldots$ However, each initialization γ_0 gives another sequence of the system over time. To really understand a self-organizing system, abstraction and simplification are a must. Therefore, we consider a simpler system ϕ with a smaller configuration space $\phi \in \Phi$ with $dim(\Phi) \ll dim(\Gamma)$. Hence, we want to find a mapping f that allows us to map the configurations γ to the simpler, abstracted configuration ϕ leading to $f : \Gamma \mapsto \Phi$. From one time step to the other, from t to $t + 1$, we consider an update rule g on the real configuration $g : \Gamma \mapsto \Gamma$, thus $g(\gamma_t) = \gamma_{t+1}$, and an update rule h for the abstracted configuration $h : \Phi \mapsto \Phi$, thus $h(\phi_t) = \phi_{t+1}$. This is summarized in Figure 3.1. Finally, the abstraction implemented by f needs to be chosen carefully such that after the updates through g and h the correct, but abstracted model, remains the right one in the next time step.

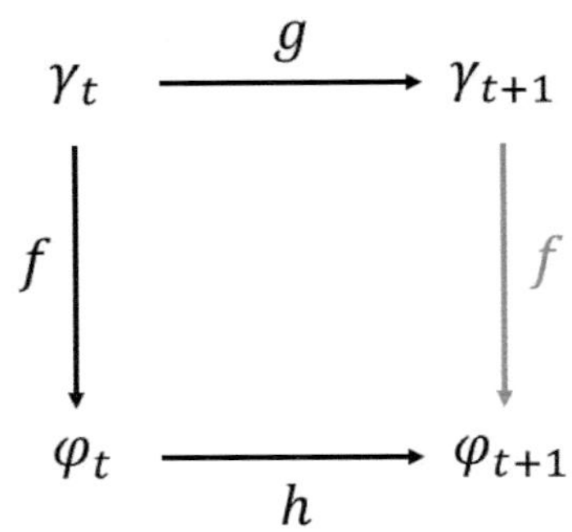

Figure 3.1: The modeling and abstraction of a self-organizing system adapted from [153].

Nevertheless, always keep in mind: "Essentially, all models are wrong, but some are useful" (George Box, 1976). In other words, all models abstract details from reality; while no model can perfectly capture its complexities, they can still provide valuable insights and guide decision-making in uncertain situations.

3.1 Modeling Principles

Generally, there is no common modeling approach for self-organizing systems or swarms of CPSs specifically [153]. Various models have been developed to work for different setups. Autonomic characteristics like self-configuration, self-optimization, self-healing, self-protecting and self-awareness schemes can be modeled by for example graph rewriting [72], L-systems [298], matrix rewriting [353], rewriting game theory [61], reaction-diffusion models [211]. In this book chapter we will not explain all the modeling approaches as they are quite diverse from application to application. We rather focus on use-case-specific models (Section 3.2 and Section 3.3) that we worked on in different projects.

When designing a model for a system, the first step is to determine which components should be represented as discrete elements and which should be modeled as continuous entities. There are four significant aspects where this decision profoundly influences the design and behavior of the model: Time, space, states of agents, and interaction. When we use a discrete micro-level model [165, 366], the behavior is usually event-driven: Each time an event occurs in the system, nodes may change their internal states and produce local output, which is sent to other nodes. This behavior can be modeled by finite automatons. Since many systems are too complex to be modeled in detail, probabilities can be used to describe the behavior. This leads to the concept of stochastic automations. In a continuous micro-level model [164], the state of each node changes continuously over time. The behavior of a node can be described by a local differential equation, which takes the input values

of other nodes into account. There are numerous modeling techniques, which have already been used [248], including

- Maxwell's demon, a model to explore the concept of entropy,
- Turing machine, a model to explore the concept of computation,
- Cellular automaton, a model for complex systems to perform discrete micro-level modeling, and
- Von Neumann's self-reproducing automation, a model to explore the logic behind self-reproduction.

Generally, a system's model is always influenced by its environment and the application it is intended for. This influence helps us to identify the most key features. When modeling an application as a self-organizing swarm, we must consider a set of potential swarm members and their characteristics. These agents can either be of the same type (homogeneous swarm), or different types (heterogeneous swarm). Additionally, we have established criteria to determine whether an agent is eligible to be a swarm member in the application. These criteria include whether a swarm member

- Is accompanied by a reasonable number of other swarm members in the system,
- Exhibits an appropriate level of abstraction for modeling,
- Can detect and respond to information from the local neighborhood and environment, and
- Is plausible and understandable, fostering trust during the modeling process for the proposed solution [336].

3.1.1 Challenges

When modeling a system as a swarm of CPSs several questions need to be addressed [336]:

1. **What are the most representative agents and the corresponding level of abstraction?**
   A swarm consists of agents that interact with the environment and with each other. Each agent serves as a digital twin, an abstraction of a real CPS. An initial step in modeling a swarm algorithm requires a decision on which entity will serve as an agent within the swarm. As described in the beginning of Section 3.1, we consider several criteria to support this decision. However, determining the appropriate level of abstraction can be challenging: For instance, in a search and rescue task (see Section 3.3 for more details), we could consider robots as agents at a higher abstraction level treating robots of the same type

as agents. Alternatively, we could choose a lower abstraction level and model individual hardware components (Central Processing Unit (CPU), memory, motor blocks, etc.) as agents.

2. **How to deal with inhomogeneities among agents?**
   Agents are inherently heterogeneous: Even minor variations in deployment can lead to differences in their reactions or behaviors, potentially resulting in unintended system outputs. For instance, in a fab setting (see Section 3.2 for more details), there is no standardized type of machine, as each can run different processes locally. Consequently, each machine has a unique set of parameters influencing its operation (e.g., furnaces use batch-processing and waiting too long to fill up the batch machine could waste time in the production process). This variability must be considered during the modeling process.

3. **How to implement the necessary local swarm communication paradigms in the specific use case with the chosen agents?**
   Communication paradigms for swarm algorithms include direct communication and indirect communication. In direct communication agents send messages to other agents. In indirect communication, also referred to as stigmergy, agents leave information in the environment for other agents to "pick up". Dependent on the level of abstraction the communication is performed literally among the agents. If the agents are already represented as digital twins in the central computer system, thus even more abstracted than only on CPS level, the communication can easily be implemented as local messages within the computer system or in local memory for stigmergy. The latter case also lightens the requirements for abstraction of the CPSs, as it is not necessary to equip, for instance, machines or products with additional local computational or communication intelligence.

4. **How to implement a self-organizing solution on top of a working system?**
   Typical technical systems already include a set of mechanisms, rules, priority classes, etc., that trigger the sequence of task execution. Therefore, it is important to decide which parts of the system should run as they are, and which parts should be abstracted away—step-by-step. For example in case of priority classes, that also grow historically, the classes could be reduced to a single one by introducing swarm-based local rules and interactions.

5. **How to validate the approach?**
   Testing and predicting the performance of a given self-organizing algorithm is difficult. Historical data that might exist for a running system, or the implementation of baseline algorithms should be used to compare with system. As the sequence of processed behavior is not deterministic, but rather stochastically differs dependent on the

initial condition and initial parameters, the validation process requires multiple simulation runs. Another challenge that arises is debugging a failed validation. While simulation runs might inform the user that the intended level of service for a given system is not achieved for some test cases, pinpointing the specific problem and rectifying it can be very challenging for a swarm system. This is because even a minor behavioral change in an agent can result in unexpected and undesired emergent behavior at the system level. Utilizing deterministic simulations with seeded random number generators can aid in the debugging process, but they do not address the overarching issue of applying a corrective solution.

3.2 Use Case: Models in the Production Plant

Scheduling in modern production plants that follow the flexible job shop principle presents a challenging, dynamic problem within the context of Industry 4.0. A typical application is the production of Integrated Circuits (ICs) in the semiconductor manufacturing industry [138]. Particularly, we consider the processing of wafers to create ICs at the so-called front end of line processing. In contrast to the high-volume production of memory ICs and CPUs, the logic and power sector manufactures fabricate specialized ICs in low-volume batches in the same plant. For this example, we consider the requirements and constraints of the leading semiconductor manufacturer Infineon Technologies AG [175]. They need to schedule between 400 and 1200 stations in their production plant producing more than 1500 different products in around 300 process steps each. The steps for a product include various processes including lithography, doping, oxidation, etching, and measuring [138].

In this use case several constraints influence the scheduling process, including the need for secondary resources, constraints from machines with equipment tooling, time couplings and batch processing. The production process involves successive steps that often loop back, requiring the same machines or machine groups to be used repeatedly. Furthermore, the production plant is required to meet various global objectives, such as maximizing machine utilization, throughput time, delivery reliability, and minimizing Work in Progress (WIP). The presence of a wide range of product diversity, coupled with the historical growth of the industrial plant, further amplifies the complexity of the system [336]. With all these constraints and requirements, we can consider the job shop scheduling problem as NP-hard [130].

Existing dispatching rules are based on heuristics and can only be used on a subset of the plant. Similarly, linear optimization methods can only cope with a subset of the plant and not with the large, and dynamic search space of an entire fab [222]. The reason is the excessive computation time needed for the calculation. Thus, these methods do not exploit the full optimization potential.

In the production plant this leads to bottlenecks that cannot be prevented, and WIP waves that will be generated. Summarized, no optimal solution for job shop scheduling has been developed so far using linear optimization, especially a solution that can be computed in polynomial time [425].

This situation paves the way for a different approach, a modeling and optimization of the production plant from the bottom-up.

3.2.1 Swarm Member Candidates

With the general problem formulation in the beginning of this Section 3.2 we identify a number of possible agents in the production plant that can act as swarm members and can be modeled in a network model [336].

Lots (L) can form swarms with lots of the same product type (homogeneous swarms), or swarms with lots of multiple product types (heterogeneous swarms). Lots follow a specific recipe through the production plant, which prescribes the order of processing steps, but not the specific machine on which to perform the next required processing step. Typically, there are multiple machines that can perform the same process. Therefore, the lots could decide which machine to prefer over the other. Additionally lots can manipulate their own priority that allows a lot to be promoted in the machine's queue. The path that is taken through the production plant by a lot is called a route. It is obvious that lots of a similar product type will share parts of their recipes. Therefore, in a stable load situation of the production plant, lots of the same and similar product types can be expected to share parts of their routes.

Machines (M) have the local information of the processes they can perform and their utilization. Given the presence of various machine types within the production plant, they can be organized into either multiple cooperating homogeneous swarms or a heterogeneous swarm with diverse capabilities. The neighborhood of each machine is determined locally and dynamically based on the recipes of the incoming and outgoing lots. Thus, the neighborhood represents not necessarily machines that are physically close, but rather those that are close according to the recipe of a lot. Machines can take decisions locally and can, if necessary, select which lot to process next by re-ordering their queue. Furthermore, they can locally communicate with other machines in their neighborhood, and could ask for lots that have their available specific processing type.

Workcenters (W), with $W \subset M$, consist of multiple related machines that could be of the same or similar type. They have attributes very similar to single machines. A workcenter can calculate when lots will be processed internally and can use the makespan information (the time between the start and the end of a lot's production) to calculate when the lots currently being processed will be finished.

Processes (P) can represent virtual swarm members. Thus, they have a view on all machines that they can potentially be run on. This also includes

the machine's current and total workload. Therefore, they can forecast the workload, the times for re-tooling at the machines, and have information on batching requirements of batch machines.

3.2.2 Network Model

For this scenario, we adopt a network model approach. The production plant is presented as a graph with the tuple $G = (V, E)$, with the notation V for the set vertices or nodes, and E for the the set of edges. We have everything that is needed for this type of modeling: Nodes that represent the machines and form a local neighboorhood together with edges that indicate a dynamic path through the production plant from one node to (many) other(s). Different from the classical graph theory, nodes and edges can be added or removed dynamically, nodes can even move (if we model them as robots). Additionally, characteristics of possible topologies (e.g., star, bus, ring) are not important as these change dynamically with the nodes and edges as well.

Considering the swarm member candidates from the previous Section 3.2.1, we apply the network model leading to a network consisting of a production plant with machines, queues, processes, lots, and recipes. The production plant $\mathfrak{P}$ consists of several sets or workcenters of machines $W^m = \{M_1^m, M_2^m, \ldots\}$, where m is the machine type. Each machine M_i^m has a queue $\mathfrak{Q}_i^m$ and every machine in a workcenter W^m can perform a process P^m. A set of lots $L = \{l_1^t, l_2^t, \ldots\}$ needs to be processed in the plant, with t as the product type. Every product type t is characterized by a recipe R^t that outlines the specific sequence of processing steps required to manufacture the product. Each lot l_n^t has the flexibility to select the appropriate machines M_i^m for each necessary process step P^m from the available options.

Based on this formal representation, the recipes can be interpreted as a directed graph $G = (V, E)$ that represents the possible connections between the machines within the plant. The nodes V correspond to the machines M_i^m, and the edges E are established between two machines M_i^m and M_j^p if there exists a lot l_n^t with a recipe R^t that includes consecutive processes P^m and P^p. A route $\mathfrak{R}$ can be defined as an ordered list of machines capable of executing the successive processes outlined in the recipe. The taken routes are a subgraph of G with $G_{\mathfrak{r}} \subseteq G$.

3.3 Use Case: Models in Search and Rescue Applications

The SAR use case envisions a heterogeneous swarm of UAVs and Unmanned Ground Vehicles (UGVs) that is deployed in a highly dynamic disaster environment. Their task is to support first responders in real-time by locating human casualities or trapped individuals and providing first aid to those found. The use case of the acpCPS swarms can be described as follows: The

swarm of UAVs cover a defined area and collectively search for victims. As soon as a UAVs discovers a victim, it switches from the coverage to a tracking task. By tracking the victim it continuously reports the current victim's location to a specific UGV. The target UGV to be informed is selected by the UAV using a specific cost function, for instance, the distance of the UGV to the victim. Finally, the UGV is assumed to rescue the victim by navigating to the location provided by the UAV. As soon as the UGV reaches the victim, it returns to its base, and the UAV rejoins the coverage task of the UAV swarm to locate other possible victims. We implemented this use case in the CPSwarm[1] project (for more details on the project, please visit [25]).

The constellation of this mission is well suited for testing swarms of CPSs and swarm behaviors because it can benefit from several swarm characteristics (see Chapter 2 for more details): The scalability of a swarm enables the addition of more CPSs while executing a mission. An example is the case where the area to be covered is larger than expected. Robustness of a swarm guarantees a mission execution in challenging environments where individual CPSs can fail. Furthermore, unlike fully centralized control, such a swarm can continue to operate even in scenarios where connectivity is limited or sparse. The swarm functions autonomously as a self-organized system composed of diverse members, making mission-critical decisions based on the demands of the dynamic environment.

3.3.1 Model-Driven Engineering of CPS Swarms

For a use case like SAR, we need a well-defined model of the CPSs and the desired local behaviors, with scalability across different abstraction levels being crucial [225]. To design a multi-scale model, we must consider both the hardware and software components of individual CPSs, as well as the overall composition of the entire swarm of CPSs. The process of modeling a CPS swarm involves making several decisions, which encompass: (i) Deployment specifics, such as determining the number, type, and placement of CPS within the swarm. (ii) Hardware considerations for each CPS, including communication technology, interfaces, onboard sensors and actuators, processing unit, and memory. (iii) Specifying the desired local behavior of individual CPS units and the overall global behavior of the CPS swarm [345].

This leads us to a model-driven engineering process that starts from modeling the CPS hardware, their local individual behavior, and their global behavior in a swarm. Finally, this approach results in behavior code that can be deployed to and executed on the CPSs. These steps rely on a hierarchically organized set of behaviors that collectively form the controller of each individual CPS. These behaviors are based on a well-defined library of behavior models, which are linked to an automated code generation process, enabling the assembly of executable code for each CPS [345]. Based on the

[1]CPSwarm website: https://www.cpswarm.eu/[Online; accessed 19-December-2024].

multi-scale modeling approach, it is possible to validate the swarm behavior. This is achieved by an abstract simulation allowing for rapid prototyping of the CPS behavior to observe the resulting and the desired swarm behavior. All results can be found in Sende et al. [345].

3.3.2 Hardware Modeling

When beginning to model a CPS we examine its architecture, focusing on the hardware components, their interfaces and interactions. For creating hardware models we utilize Systems Modeling Language (SysML) [15], an extension of Unified Modeling Language (UML), that focuses on modeling systems with block diagrams. To advance the modeling of CPS swarms, we enhanced the Block Definition Diagram (BDD) and Internal Block Diagram (IBD) by incorporating a CPS swarm profile [335]. It includes three types of diagrams [345]:

- **Swarm composition diagram:** We developed a swarm composition diagram that extends the BDD, enabling the modeling of the entire swarm. The diagram details the specific types and quantities of CPSs used within the swarm.

- **Hardware composition diagram:** This diagram is designed to extend the BDD that allows the modeling of individual hardware components installed utilized by the CPSs. The hardware composition diagram specifies the types of hardware components in use and defines their inputs, outputs, and other parameters.

- **Swarm member internal diagram:** We designed the swarm member internal diagram to extend the IBD to model the internal structure of a single CPS. This diagram specifies how the behaviors can interact with the environment or other CPSs by defining sensors, actuators, and communication interfaces and their parameters. On the one hand, sensors are inputs to collect information from the environment, on the other hand, actuators are outputs to interact with the environment. A communication interface, however, can provide both, inputs and outputs to the swarm behaviors and thus, enables coordination between CPSs.

3.3.3 Behavior Modeling

The behavior models describe how the individual CPS behaves, also when interacting with the environment and other CPSs in the modeled swarm. This way, the behavior models define the software components for each CPS. The individual connected behaviors are the steps each CPS must execute locally to collectively achieve the global mission that envisioned by the swarm system's designer. The design of local behaviors for the individual swarm members is a difficult task, as the emergent global swarm behavior is not easily predictable [1]. Modeling the behavior on different abstract levels supports

the process to generate the desired global behavior as it can be executed on different levels of realism and detail [345]. With this bottom-up approach, we can iteratively refine the individual local behaviors until we achieve a global swarm behavior.

Behavior State Machines

Once a complete mission is defined, we typically identify a set of distinct behaviors for each CPS in the swarm to execute, ensuring the various tasks of the mission are accomplished. These individual behaviors are treated as atomic in the modeling process, making them simple to describe and implement. By combining these simple behaviors, we can create more complex behaviors to accomplish intricate missions. A widely used yet powerful approach is to employ Finite State Machines (FSMs), where each state represents a simple behavior, and the transitions between states depict the shift from one behavior to another. Thus, the entire FSM design describes a complex behavior. During operation, each CPS runs an FSM, consistently remaining in a defined state, though this state may differ among CPSs. Hence, complex swarm behaviors can emerge where CPSs play different roles based on their interactions [345].

The transitions between the behaviors described as states in a FSM are triggered by events. These events can either originate locally, such as from sensor outputs or the behavior itself, or from the environment, like communication activities between CPSs or external commands from a global control station. The exchange of events between CPSs facilitates the coordination of local swarm behaviors. Events are processed locally and autonomously by each CPS individually as specified in their behavior FSMs. Through the exchange of events, CPSs can exert an impact on each other's behavior changes. These events are characterized by a distinct Identifier (ID), a timestamp, and a unique sender ID. Additionally, events can carry associated data, which is transmitted between behaviors.

The UML behavior FSMs [68] serve as the foundation for the behavior FSM model. Simple behaviors are represented by simple states, while composite or submachine states are used to model complex behaviors. Composite states enable the modeling of one state using another FSM, while submachine states facilitate the encapsulation of generic FSM that can be utilized in multiple states. In the realm of CPS swarm behaviors, we introduce four distinct behavior types, each denoted by a specific color code as illustrated in Figure 3.2.

- Swarm behaviors, represented by the green elements, are simplistic behaviors that execute specific swarm algorithms. These algorithms give rise to emergent swarm behaviors, such as aggregation or exploration.

- Swarm functions in blue are simple behaviors that carry out a singular function encompassing the interaction between CPSs. Examples include task allocation or the exchange of position information.

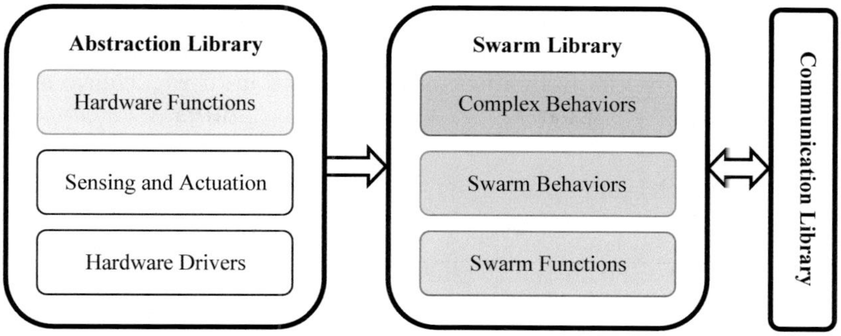

Figure 3.2: The behavior library structure [345].

- Hardware functions, depicted by the yellow elements, are straightforward behaviors that perform singular functions involving hardware interaction. Examples include moving to a designated location or controlling an actuator.

Each behavior is given a distinct name and a brief description, and is characterized by its behavior type as well as its respective inputs and outputs.

It should be noted that in related literature, such a granular differentiation between various behavior types is not commonly found. For example, swarm behaviors and swarm functions are termed collective behavior in [45] and basic swarm behavior in [337]. In [124], the term "constituent behavior" is employed to refer to any kind of simple behavior, including hardware functions. However, we intentionally introduce the distinction between behavior types to promote a structured and organized design of the behavior FSMs. This is facilitated through the use of hierarchically nested states, as outlined in the UML standard. This enables the abstraction of behavior details at higher levels. For instance, an aggregation behavior can be implemented independently of the method employed for exchanging position information between CPSs. Events can be utilized to trigger a state change within the sub FSM or cause the higher-level behavior to terminate, thus also concluding the currently running sub-behavior.

Behavior Libraries

To enable frequently recurring behaviors and functionalities to be defined only once, the behaviors are organized into libraries. The library structure we propose is illustrated in Figure 3.2, with each color corresponding to a different behavior type as previously defined. The libraries are organized based on the level of hardware abstraction, which facilitates the separation of concerns. They comprise software artifacts that are modeled as states within the FSMs. The Swarm Library is hardware-independent and contains the complex behaviors modeled as FSMs, in addition to swarm behaviors and swarm functions. It utilizes a Communication Library, which provides a

communication interface between CPSs. The Abstraction Library abstracts the hardware specifics and provides functions related to hardware, as well as functionalities to access sensors and actuators using hardware-specific drivers.

The Swarm Library consists of swarm behaviors executed by individual CPSs, resulting in the global behavior of the swarm. They are platform-independent and can be reused across different types of CPSs. The library is divided into three sub-libraries based on the previously introduced swarm behaviors. Firstly, the Complex Behaviors Library contains FSMs that model complex, high-level mission behaviors such as SAR, and are defined as UML composite or submachine states. Secondly, the Swarm Behaviors Library comprises individual swarm algorithms that exhibit emergent behavior. These swarm algorithms are typically based on biological inspiration or are generated automatically, such as through evolutionary optimization. Examples include flocking, phototaxis, or collective transport, and they are defined as UML simple states that are used in the complex behavior FSMs. Lastly, the Swarm Functions Library includes simple swarm-related tasks that do not lead to emergent behavior, but rather enable the functioning of the swarm behaviors. Examples include the exchange of position information, task allocation, or computation of the average velocity of the swarm. They are defined as UML simple states to be used in the complex behavior FSMs.

The Communication Library provides a set of communication services to the swarm, which are built on top of an arbitrary network interface. These services encompass various functionalities, such as transmitting telemetry data from the CPSs to the command and control station, facilitating the exchange of events between CPSs, as well as between CPSs and the command and control station, and enabling remote access to parameters of the CPSs within the swarm. This library was developed as part of this work and is available on GitHub [297].

The Abstraction Library provides a set of functions and interfaces that abstract away the hardware specifics of the CPSs. It allows the development of high-level routines from an application-oriented perspective, raising the level of abstraction from a platform-dependent point of view. This enables developers to focus on describing how the CPSs should behave in order to complete a high-level task or reach an application-specific goal, without being concerned with the underlying hardware implementation details. This is achieved by providing a set of CPS-specific libraries in order to access platform-specific information of a acCPS in a standard and coherent way.

The Abstraction Library is designed to provide a higher-level interface for accessing the hardware of the CPSs, allowing developers to focus on high-level tasks and goals rather than low-level hardware details. The library is composed of three layers, with each layer adding a level of hardware abstraction. The bottom layer, Hardware Drivers, includes all the drivers for sensors and

actuators that are mounted on the CPSs. The middle layer, Sensing and Actuation, is responsible for providing sensor information and controlling the CPSs using their actuators. The Hardware Drivers layer establishes a direct connection with the hardware, while the Sensing and Actuators layer employs software to offer an initial level of abstraction by implementing intricate functionalities needed by the higher layer. At the topmost layer, the Hardware Functions layer encompasses a collection of high-level functions that represent sophisticated routines executed by a CPS, involving a combination of sensors and actuators. These functions are used to define the states of the FSMs as UML simple states.

3.3.4 Code Generation

The responsibility of the code generator is to convert the FSM models into executable code that can be deployed on the CPSs. The generation process is based on templates, which are suitable for structures that are schematic and repetitive. With template-based generation, a basic set of templates is defined, which is then populated with data extracted from the algorithm specification.

The templates consist of a static part that appears in the output as-is and a dynamic part that is replaced by input data using a template meta-code. The meta-code includes directives that are processed by a template engine along with the input data to produce the final source code. To target different runtime platforms, different templates are used. The code generator uses a set of templates written in the Velocity Template Language (VTL) to generate executable code.

The code generator processes the data in the form of FSM models of complex behaviors, which are translated into State Chart XML (SCXML) files. These files contain all the necessary information for the code generator, including the type of functionality used to select the appropriate template and the definition of the Application Programming Interface (API) executed in a given state.

SMACH is a Python-based project that provides a framework for the implementation and execution of FSM-based algorithms. The code generator, which is part of this project, uses the SCXML file and a set of templates to produce Python code that implements the designed FSM. This code generator is available on GitHub [296].

3.3.5 Modeling on the Example of the SAR Use Case

In this section, we will show how the formal approaches proposed earlier can be applied to a heterogeneous swarm of CPSs. We will describe how we model the hardware and software components that were used in our experimental evaluation. Our main objective is to demonstrate the feasibility of deploying software onto CPSs based on the models created using formal methods. The diagrams were created using the modeling tool Modelio. Subsequently, the data was exported for additional processing by the code generator, utilizing

the CPSwarm Modeler module. This module can be found on GitHub [295], providing the necessary functionality for the generation of code based on the exported data.

We have selected the SAR use case to showcase the formal approaches proposed in this section. CPSs have various applications in such missions, such as creating a situational overview, providing logistic support, serving as a repeater or surrogate for other CPSs, and clearing debris, as reported in Murphy et al. [262]. To make the SAR use case more manageable for deployment on prototype platforms that do not necessitate specialized hardware, our focus is primarily on two tasks: Locating human casualties or individuals trapped in the disaster site and providing initial medical assistance. To accomplish this, we implement a heterogeneous swarm comprising both UAVs and UGVs. The mission of the heterogeneous CPSs swarm can be described as follows: The UAVs cover a designated area to search for victims, while the UGVs remain inactive. Once a victim is detected by a UAV, it switches to tracking mode and continuously tracks the victim's position, which is communicated to the UGVs. An arbitration process is initiated by the UAV to select the most suitable UGV to reach the victim, based on a cost value provided by the UGV, such as the distance to the victim. The chosen UGV navigates to the victim using the received position information from the UAV, and once the victim is rescued, the UGV returns to its starting position while the UAV resumes coverage to search for a new victim.

As an example, we will focus on modeling the UAVs, including their hardware models and behavior implementation based on Robot Operating System (ROS) [300], while omitting the simpler UGV models for brevity. To simplify the SAR use case, we will refer to victims as targets.

Hardware Models

The authors of this work have custom-made the UAV platform as prototypes. The UGV platform is based on an off-the-shelf RC truggy. The UAV prototype platform is represented by an extended SysML IBD model, as depicted in Figure 3.3. The model defines the inputs such as sonar range finders, camera, and Ultra Wideband (UWB) localization. The outputs for controlling the locomotion and a communication interface for inter-CPS communication and communication with the command and control station are also specified. Further information on this modeling process can be found in [332].

Behavior Implementation

To represent the behavior of the UAVs, we utilize a two-level FSM hierarchy denoted as $H = \{L_1, L_2\}$. This hierarchy effectively models the swarm behaviors using a collection of swarm and hardware functions sourced from various libraries. The implementation of these behavior libraries is built upon ROS, an open-source framework specifically designed to simplify the creation

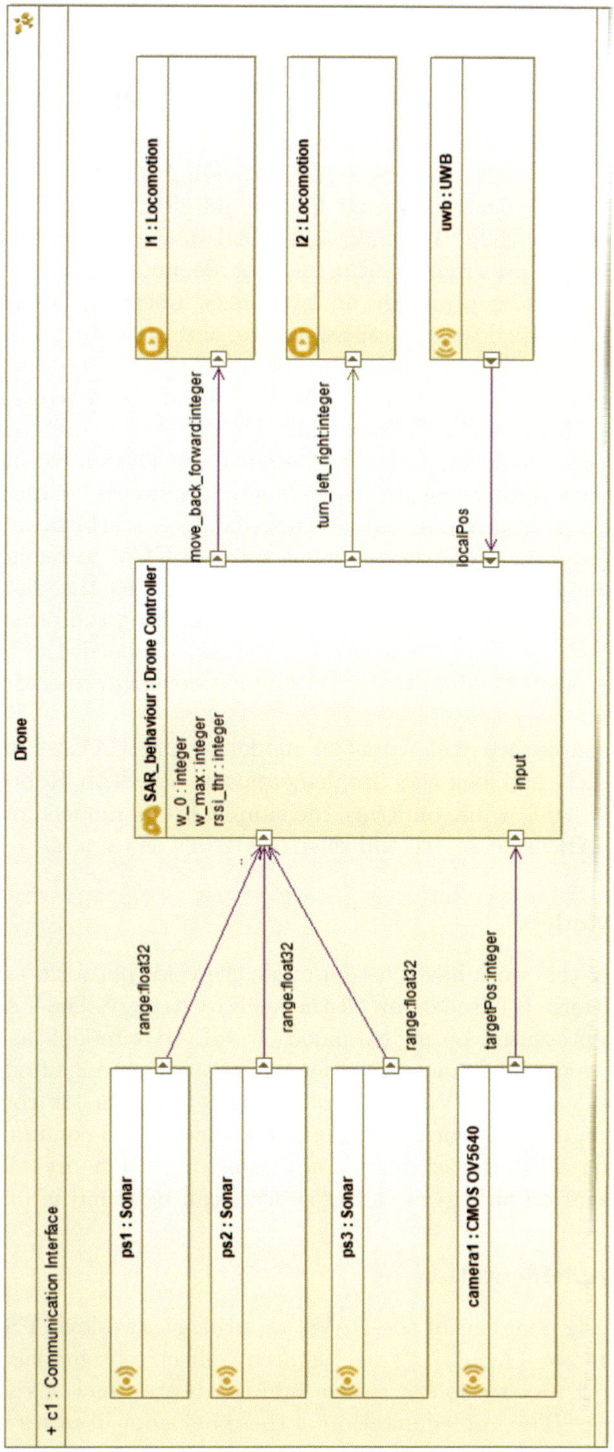

Figure 3.3: The hardware model of the UAV prototype [345].

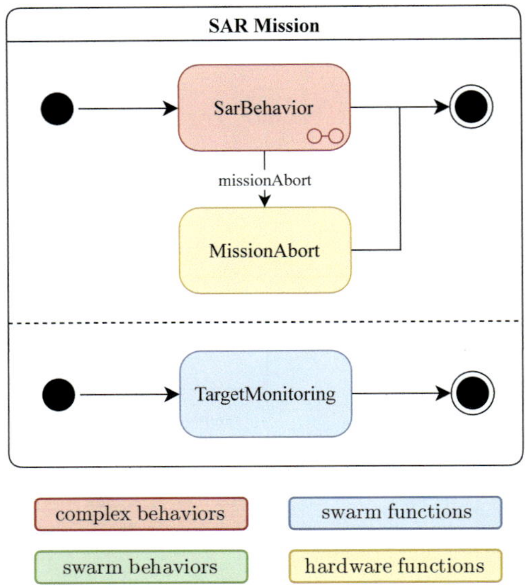

Figure 3.4: The UAV SAR mission processes [345].

of robot software. ROS has gained widespread adoption in the robotics community due to its exceptional flexibility and extensive hardware platform support, making it the de facto standard in the field.

The first level L_1 in the hierarchy defines the parallel processes that are executed on each UAV, as depicted in Figure 3.4. The SarBehavior state models the actual SAR behavior required to complete the mission and is itself a complex behavior modeled as a L_2 FSM. The *missionAbort* event allows for the immediate termination of the SarBehavior, along with any currently running behavior in the corresponding L_2 FSM. Therefore, an event broadcasted by the command and control station to all CPSs can stop the entire mission. There are various swarm and hardware functions executed in parallel, which are specified in detail in Tables 3.1 and 3.2, respectively.

The UAV FSM hierarchy L_2 consists of the second level, which models the behavior of the UAV during the SAR mission. It is depicted in Figure 3.5. When a UAV is powered on, it starts in the Idle state. Upon receiving the *launch* event from the command and control station, the UAV transitions to the TakeOff state and ascends to the designated altitude. Once the desired altitude is reached, the UAV enters the Loitering state and remains stationary. The actual SAR mission starts when the *missionStart* event is received. The coverage state is the initial state of the SAR behavior, during which the UAV searches for targets. If a target is detected, the TargetMonitoring state triggers the *targetFound* event, leading the UAV to the SelectRover state. During this state, the UAV communicates with all available UGVs and selects the closest one to move towards the target. As the selected UGV moves towards the target, the UAV performs the Tracking behavior to keep track of the target's

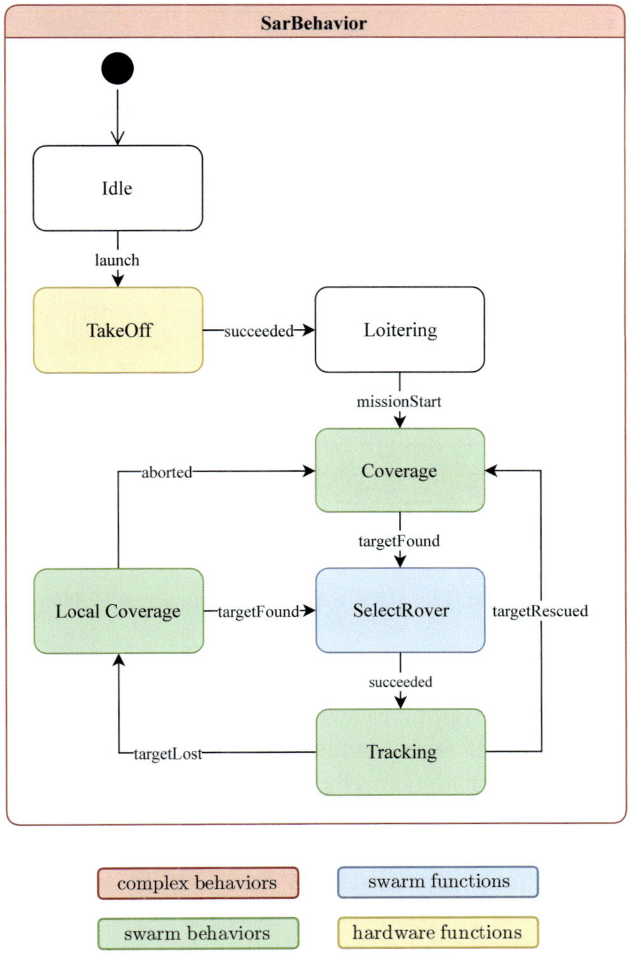

Figure 3.5: The complex SAR behavior of the UAVs [345].

location until it is reached by the UGV. It informs the UGV about changes in position through the *targetUpdate* event. If the *targetRescued* event is received from the UGV, the target is safe, and the UAV restarts the "Coverage" behavior to search for other targets. If the target is lost before being rescued, the UAV enters the LocalCoverage state, which allows it to circle around the last known target position to find the target again. If the target is found, the UAV selects the most suitable UGV again. If the target cannot be found, the UAV restarts the "Coverage" behavior.

The SAR mission's FSMs utilize behaviors and functions from the Swarm Library and the Abstraction Library. Table 3.3 and Table 3.1 present the swarm behaviors and functions used in the SAR mission, respectively. The complete libraries are available as ROS stacks, the behaviors [341] as well as the functions [342], and their details can be found in the links provided in

Table 3.1: The swarm functions used in the SAR mission as part of the Swarm Library [345].

Behavior	Input	Output	Description
TargetMonitoring	camera footage	IDs of targets	Manages targets being detected by the swarm. It uses the on-board camera to detect targets and exchanges information about targets with the other CPSs in the swarm.
SelectRover	target ID, target position	UGV ID, target ID, target position	Assigns the closest idle UGV for rescuing a target using the *targetAssigned* event.

Table 3.2: The hardware functions used in the SAR mission as part of the Abstraction Library [345].

Behavior	Input	Output	Description
MissionAbort	-	-	Lets a UAV land.
TakeOff	altitude	-	Lets a UAV lift off.

Table 3.3: The swarm behaviors used in the SAR mission as part of the Swarm Library [345].

Behavior	Input	Output	Description
Coverage	coverage area boundaries	target ID, target position	Lets a UAV fly over the mission area in search of targets. It terminates once it finds a target.
Tracking	target ID	target position	Lets a UAV keep track of a target that has been found. It informs other CPSs about position changes using the *targetUpdate* event.
LocalCoverage	last known target position	target ID, target position	Lets a UAV search the local neighborhood for a lost target. This exploits the fact that a target has a high probability of still being close to the current position of the UAV.

the footnotes. The hardware functions of the Abstraction Library utilized in the SAR mission are listed in Table 3.2. There are other modules pertaining to different levels of the Abstraction Library released on GitHub. For more information, please refer to the repositories of the Hardware Functions [343] and Sensing and Actuation [344] libraries, as specified in the footnotes. The events that trigger state changes in the behavior are summarized in Table 3.4.

Table 3.4: Events used in the SAR mission [345].

Identifier	Data	Sender
launch	-	command and control station
missionStart	-	command and control station
missionAbort	-	command and control station
targetFound	target ID, target position	swarm member
targetUpdate	target ID, target position	swarm member
targetLost	target ID, target position	swarm member
targetRescued	target ID	swarm member
targetAssigned	target ID, UGV ID	swarm member

3.4 Use Case: Models in Edge Computing

The shift toward local processing at the edge brings critical benefits that address future computing challenges, including enhanced security, improved reliability, reduced latency, and lower energy consumption. Managing this edge infrastructure, often called the edge continuum, creates a dynamic and adaptable computing environment.

In this use case, we examine a network of Edge Micro Data Centers (EMDCs) within the edge continuum (see Figure 3.6), where intelligence is distributed across the nodes, creating a decentralized environment. This distribution enhances the edge's autonomy and granularity in local decision-making within a regional context, reducing dependence on a central coordination point. This is particularly crucial for real-time applications such as autonomous driving or the monitoring and control of smart grids. The stability and performance of edge infrastructure are increasingly challenged by stringent requirements for latency and autonomy, distribution across multiple locations, limited local size, multi-tenancy, the involvement of multiple operators, and local management, with components operating concurrently and asynchronously.

These challenges are further intensified by the rapid growth of i) connected devices and their data exchange capabilities, ii) intelligence embedded in edge devices, iii) the breakdown of monolithic applications into smaller components, and iv) the scale, speed, and complexity of interactions among devices in a zero-trust environment. As a result, orchestrating edge and cloud interactions, particularly in resource allocation, workload scheduling, and data management, is becoming increasingly complex (see Figure 3.6 for details on the architecture [334]).

To tackle these challenges, we propose an approach centered on autopoietic systems—self-organizing, self-regulating, and self-repairing. AI-driven optimization methods, such as swarm intelligence, have been successfully applied in cloud environments and are central to our design. Our framework uses swarm agents to represent demand and supply entities. Demand swarm agents optimize operations at the pod level, while supply swarm agents manage tasks such as workload placement and caching within

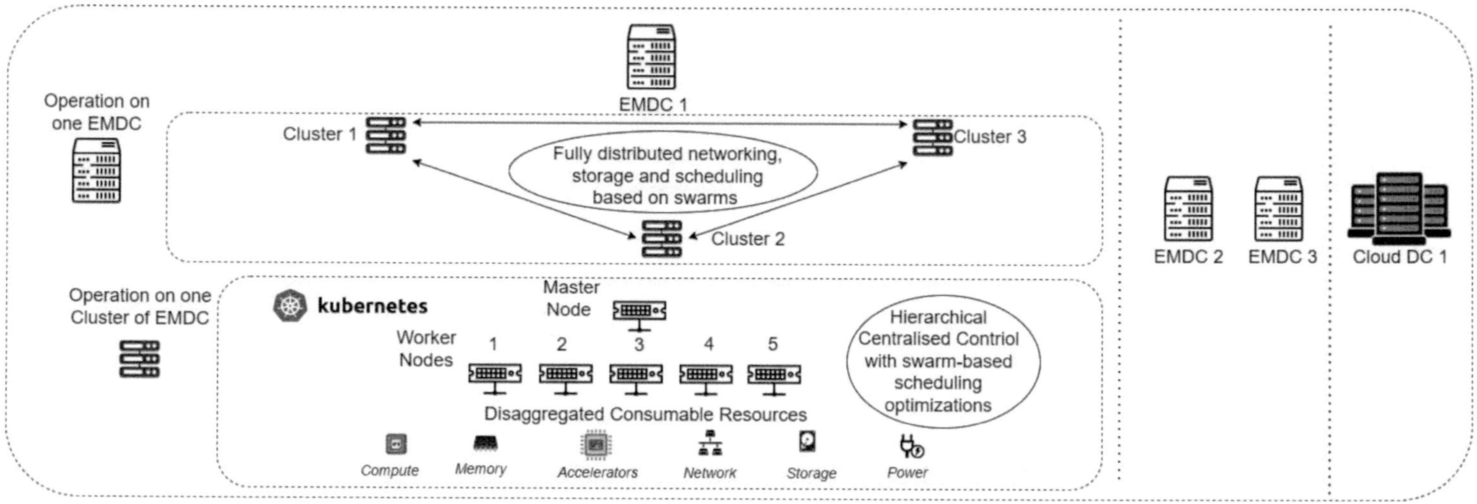

Figure 3.6: Schematic architecture demonstrating the inter-edge resource allocation in clusters: Nodes, pool of resources and the overall edge-cloud interaction [334].

EMDCs. Communication between these agents is facilitated by synthetic hormones, with supply agents detecting these signals to make informed resource allocation decisions. Inspired by ant foraging behavior, the ant algorithm is applied to optimize workload-node assignments, using pheromone trails to guide future decisions. For further information, refer to Schranz et al. [334].

3.4.1 Modeling Agents in the Edge Continuum

Our agent-based approach employs two distinct types of swarm agents: Demand and supply swarm agents. These agents work collaboratively within an EMDC environment, managing tasks such as pod placement, storage management, and caching optimization. The model represents an edge continuum comprising resources, queues, pods, and processes.

Demand Swarm Agents

Applications are decomposed into a set of services $\mathfrak{S}$, each represented by a collection of related pods $P^s = \{p_1^s, p_2^s, \ldots\}$, where s denotes a specific service. Each service s is characterized by a set of resources R^s detailing the processing steps required to execute the individual pods. Each pod p_j^s has the flexibility to select from suitable nodes N_i^n to execute the necessary process steps P^r.

Demand swarm agents operate at the pod level within the EMDC architecture, equipped with a pod-level description that contains contextual information about the workload, such as workload type, performance requirements, data dependencies, and latency constraints. These agents leverage this information to make intelligent decisions regarding workload placement and resource allocation, enhancing the overall efficiency of the system.

Supply Swarm Agents

The EMDC $\mathfrak{E}$ encompasses various sets of nodes, each with different resource types $N^r = \{N_1^r, N_2^r, \ldots\}$, where r represents the resource type(s). A typical EMDC node may offer multiple resource types, whereas a node with a single resource type might represent, for example, a CPU within a larger resource pool. In this context, we consider resources such as CPU, FPGA, RAM, and NVMe, each with specific capacities and an associated queue Q_i^r.

Supply swarm agents operate at the node level within the EMDC, representing individual nodes characterized by their available resources and respective capacities. These agents play a crucial role in the efficient allocation of resources to incoming workloads, ensuring optimal processing and resource utilization.

Agent Collaboration and Self-Organization

Collaboration between demand and supply swarm agents is facilitated through swarm intelligence algorithms. Demand swarm agents autonomously identify the most suitable nodes for workload placement, while supply swarm agents select the optimal workloads to process based on their available resources and capacities. This collaborative, self-organizing decision-making process allows the system to dynamically allocate workloads to nodes, optimizing processing performance, reducing latency, and maximizing resource utilization.

Our agent-based model is designed to exhibit autopoietic characteristics, promoting self-organization, regeneration, and regulation within the edge continuum. Through continuous interaction and adaptation to varying workloads and resource conditions, demand and supply agents foster emergent behaviors that enhance the system's resilience and operational efficiency.

3.4.2 Challenges in Modeling Agents for EMDCs

Agent-based modeling of an EMDC presents several challenges that must be addressed during the modeling process.

Pool of Resources

Beyond the nodes in an EMDC, we consider a pool of resources as an innovative addition to the current definitions of the edge continuum. This enhancement allows for the allocation of individual resources for pod processing, separate from the traditional node-based processing capabilities, which consist of multiple resources. The resource pool is integrated into the EMDC and can be accessed by edge or edge-cloud management as needed. This approach helps prevent resource limitations, minimizes latency, and ensures the stability of other pods' performance by protecting their assigned resources from being exhausted. Currently, technologies like Compute Express Link (CXL) are being integrated into CPUs (by Intel and AMD), memory, and storage components (e.g., by Samsung), with PCIe switches anticipated by 2025. Aside from hardware development, the primary challenge lies in orchestrating these resource pools effectively. Current hub-and-spoke orchestration mechanisms are insufficient for managing such ad hoc configurability.

Application Types

Different services correspond to various application types, each with unique response time requirements:

- Long-Running Applications (LRAs): These applications instantiate long-standing pods to support iterative computations in memory or continuous request-response cycles. Examples include processing frameworks like Storm [367], Flink [121], and Kafka Streams [189]; latency-

sensitive database applications like HBase [157] and MongoDB [254]; and data-intensive in-memory computing frameworks like TensorFlow [378].

- Batch Processing: This method is typically employed when large volumes of data need to be processed at once, with results stored for later use. Batch processing can occur on a scheduled basis or at regular intervals and includes two types: Regularly recurring requests and opportunistic requests with minimal or no SLA (Service Level Agreement) requirements.
- Stream Processing: This approach handles large data volumes that require real-time processing.

Future workloads are expected to become even more complex, with LRAs, batch processes, and stream processes increasingly interconnected. This will complicate the task of categorizing applications and fine-tuning their corresponding agents.

Relationships among Pods

Demand swarm agents represent pods P^s derived from a specific service s, and these pods may have various interdependencies. Pods may need to be processed in parallel or have dependencies on one another. When a pod underperforms, the current system may generate additional pods to meet the required response times for the service s. However, these interrelationships are not currently considered in scheduler and orchestration optimization. For instance, strategically placing interacting services in proximity can significantly enhance performance:

- Service Clustering: For services comprising multiple microservices that frequently interact, placing microservices within the same region can improve performance.
- Database Proximity: Pods with heavy database dependencies should be positioned close to the database to minimize latency and enhance overall performance.

Addressing these challenges involves refining the scheduling and orchestration processes to leverage these pod relationships, thereby optimizing system performance and efficiency.

Chapter 4

Engineering Swarm Behavior

Designing behavior in CPS swarms presents a significant challenge. We've placed a strong emphasis on distinguishing between micro-level and macro-level modeling of behavior, as detailed in Section 2.1. The mission of the swarm is primarily defined at a global macro level. To accomplish this mission, we must translate local rules into executable code on controllers and processing units within each CPS, incorporating hardware-related CPS functionalities such as sensors and actuators.

4.1 Basic CPS Swarm Behavior

Figure 4.1 provides a taxonomy of swarm behaviors, based on the classification proposed by Schranz et al. [337]. In the subsequent sections, we'll start by providing a comprehensive overview of this taxonomy. For an exhaustive description of the pre-existing categories, we kindly direct interested readers to [45, 337].

Spatial Organization

These behaviors enable robots in a swarm to navigate their environment, enabling them to organize themselves or objects spatially.

- **Aggregation** causes individual robots to converge in a specific area of the environment, enhancing spatial proximity among swarm members and encouraging further interactions.
- **Dispersion** operates in contrast to aggregation, as it drives individual robots to spread across the environment. Dispersion can be

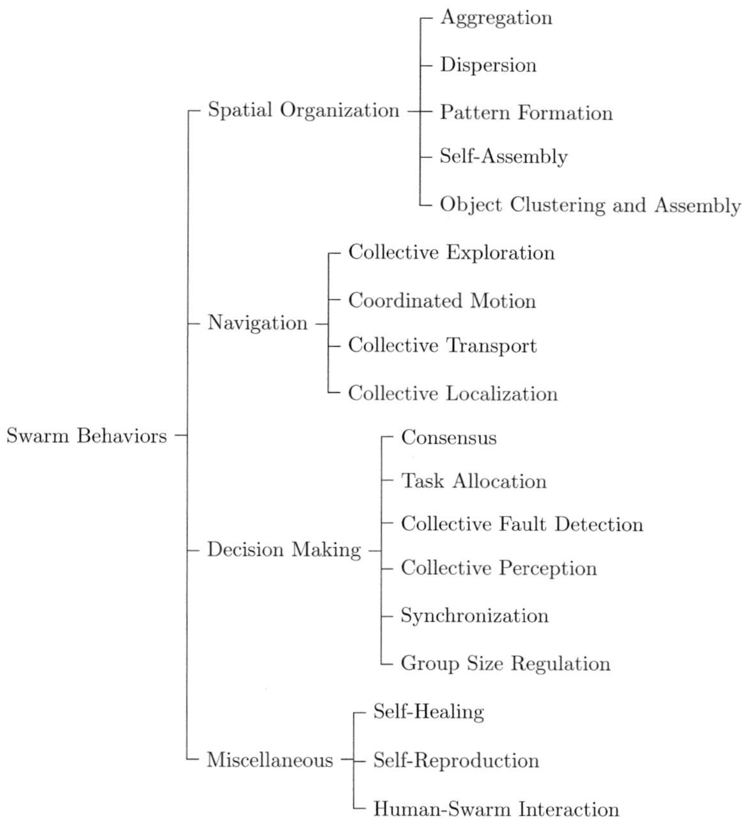

Figure 4.1: Taxonomy of swarm behaviors (extended from [337]).

advantageous in scenarios where spreading out allows for more extensive coverage of the environment or helps avoid congestion and overcrowding in certain areas.

- **Pattern formation** orchestrates the arrangement of the robot swarm into specific shapes. A notable instance is chain formation, where robots align in a line, often to establish multi-hop communication between distant points.
- **Self-assembly** interconnects robots to establish structures, either through physical connections or virtual communication links. An interesting variant is morphogenesis, where the swarm autonomously transforms into a predefined shape, exhibiting remarkable adaptability.
- **Object clustering and assembly** empowers the swarm of robots to manipulate spatially distributed objects, crucial for various construction processes, allowing for the efficient clustering and assembly of objects within the environment.

Navigation

These behaviors facilitate the coordinated movement of a robot swarm within an environment:

- **Collective exploration** guides the swarm of robots to navigate cooperatively through the environment, serving various purposes such as gaining a comprehensive situational understanding, searching for specific objects, monitoring the surroundings, or establishing a robust communication network.
- **Coordinated motion** orchestrates the movement of the robot swarm in a structured formation, which can take on well-defined shapes like a straight line or adopt more flexible arrangements, such as flocking, enhancing their collective mobility and adaptability.
- **Collective transport** empowers the robot swarm to collaboratively transport objects that exceed the weight or size capabilities of individual robots. It enables the efficient movement of bulky or heavy items by leveraging the combined strength and coordination of the swarm.
- **Collective localization** facilitates the robots within the swarm in determining their precise position and orientation relative to one another by establishing a local coordinate system across the entire swarm.

Decision Making

These behaviors enable robots within a swarm to reach a unified decision on a given matter:

- **Consensus** empowers individual robots in the swarm to reach agreement or convergence towards a single, shared choice among multiple alternatives, fostering cohesive decision-making.
- **Task allocation** dynamically assigns emerging tasks to individual robots in the swarm, aiming to optimize the overall performance of the swarm system. When robots possess diverse capabilities, tasks can be distributed accordingly to enhance system efficiency.
- Within the swarm of robots, **collective fault detection** identifies deficiencies in individual robots, enabling the identification of outliers or deviations from the desired swarm behavior, often attributed to hardware failures.
- **Collective perception** combines locally sensed data from robots in the swarm into a comprehensive view, facilitating informed collective decision-making. This behavior allows the swarm to reliably classify objects, allocate the appropriate number of robots to specific tasks, or determine optimal solutions to global problems.

- **Synchronization** aligns the frequency and phase of oscillators among robots in the swarm, ensuring a shared understanding of time. This synchronization enables robots to perform actions synchronously, contributing to coordinated and time-sensitive activities.
- **Group size regulation** empowers robots in the swarm to form groups of desired sizes. If the swarm's size exceeds the specified group size, it autonomously divides into multiple groups by simultaneously maintaining optimal group cohesion and function.

Miscellaneous

There are additional behaviors exhibited by swarm robots that do not fall into the aforementioned categories:

- **Self-healing** enables the swarm to recover from faults stemming from individual robot deficiencies. The objective here is to minimize the impact of robot failures on the overall swarm, thus enhancing its reliability, resilience, and performance. This concept aligns with the collective fault detection mentioned before.
- **Self-reproduction** empowers a swarm of robots to either generate new robots or replicate the patterns created by multiple individuals within the swarm. The aim is to enhance the swarm's autonomy by eliminating the necessity for human engineers to manually create new robots.
- **Human-swarm interaction** allows humans to control the swarm robots or receive information from them. This interaction can occur remotely, such as through a computer terminal, or in close proximity within a shared environment, facilitated through visual or acoustic cues.

4.2 Use Case: Collective Motion

Collective motion in the form of a flocking behavior, is a natural phenomenon seen in bird flocks, fish schools, and molecule movement [52]. These organisms coordinate their movement to enhance performance, such as fish evading predators and maintaining group structure or birds flying in energy-efficient V-formations. Inspired by this, various algorithms have been derived from biological examples or developed newly to address challenges in swarm robotics. Overall, these studies address how large groups of simple, small, often identical robots can collectively perform complex tasks that single robots cannot. A self-organized collective motion occurs when a swarm of robots moves cohesively, using local interactions to exchange essential information like positions, velocities, or angles.

Reynolds [311] established the foundation for modeling flocking behavior in artificial systems using so-called Boids agents. The model achieves cohesion through attraction between robots and prevents collisions via repulsive

interactions, while alignment ensures the swarm moves in the same direction at the same speed. This model assumes that each robot knows its neighbors' headings in a noise-free environment. Another method, the Self-Propelled Particles (SPP) approach, uses principles from statistical physics to achieve flocking in robots. The SPP model uses virtual attractive and repulsive forces between swarm entities to achieve self-organized collective motion [395]. There are two main approaches to reach flocking behavior: The alignment rule and the position-based rule. The alignment rule involves sharing angle information among agents to agree on their headings [394], but it requires sophisticated measurement and communication tools for reliable orientation sharing. The position-based rule, which only requires agents to detect their neighbors' positions, is a more cost-effective choice. Swarm cohesiveness and alignment are achieved using just relative position data [116]. However, it should be mentioned that, although these algorithms are cost-efficient, relying solely on relative position data results in slower response times and reduced robustness.

The SPP concept was utilized in the Standard Vicsek Model [394], an early method for flocking. This model, designed to study the effects of noise and particle size on order transitions, yielded promising results with a velocity alignment rule, where each particle adjusts its trajectory to align with the average heading of its neighbors. While velocity alignment proved useful, Couzin et al. [73] introduced another model for three-dimensional environments, considering three interactions—attraction, alignment, and repulsion—in distinct zones to simulate animal group behavior transitions.

These foundational flocking models have inspired numerous studies in swarm robotics. For example, Cucker and Smale [74] proposed a method where informed agents direct a uniformed swarm using the velocity alignment rule. This model updates agents' velocities based on a Laplacian framework, regulating velocity differences with neighbors, and inspired real-world experiments [385]. However, these frameworks require exchanging both velocity and alignment data, demanding significant onboard processing, which is challenging for small robots with limited power and processing capabilities.

Several studies have focused on implementing collective motion without relying on orientation information. For example, velocity alignment is achieved implicitly through pairwise repelling forces [244]. Other approaches include using position-based attraction and repulsion [370], inelastic collisions between isotropic agents [150], and various collective control strategies [118]. These methods avoid orientation sharing and offer distinct advantages.

Therefore, reducing the information exchanged between robots can minimize hardware complexity and cost. Additionally, studies show that alignment without explicit orientation exchange lowers the swarm's energy consumption. Ferrante et al. [116, 117] developed a similar approach with the Active Elastic Sheet (AES) model, which uses elastic interaction based on relative position for collective motion. The AES model's performance is well-documented, addressing factors like network architecture[386], robustness against measurement noise [427], and behavior under external force to guide the swarm [305]. To explain the AES model in more detail, consider a swarm

system of N robots moving in a two-dimensional arena. The motion of the ith robot is determined by attraction-repulsion forces from its closest neighbors. The positions $\vec{x_i}$ and orientations θ_i of the robots can be calculated mathematically, as described in [116, 117]. The robots are deployed in an ideal environment with negligible noise in the original signals. Thus, it is possible to eliminate disturbance terms and present a modified model as shown in Equation (4.1):

$$\dot{\vec{x_i}} = (v_0 + \alpha \ \vec{F_i}.\hat{n_i})\hat{n_i} \ , \ \dot{\theta_i} = \beta \ \vec{F_i}.\hat{n_i}^{\perp} \ , \ \hat{n_i} = \left[\ \cos(\theta_i) \quad \sin(\theta_i) \ \right]^T \ , \tag{4.1}$$

where, α and β are inverse transitional and rotational parameters, and v_0 is a biasing speed. $\hat{n_i}$ is a unit vector parallel to the heading direction of the robot i, and $\hat{n_i}^{\perp}$ is a unit vector perpendicular to it. The interactions between robot i and its neighbors s_i will generate a linear force $\vec{F_i}$ to maintain the distance within the swarm. This force can be obtained using Equation (4.2):

$$\vec{F_i} = \sum_{j \in S_i} -\frac{k}{l_{ij}}(|\vec{r_{ij}}| - l_{ij})\frac{\vec{r_{ij}}}{|\vec{r_{ij}}|} \ , \ \psi = \frac{1}{N}\left\| \sum_{i=1}^{N} \hat{n_i} \right\| \ . \tag{4.2}$$

Here, $\frac{k}{l_{ij}}$ is the spring constant, and l_{ij} is the natural length connecting robots i and j. The distance between the ith and j^{th} robots is represented as $\vec{r_{ij}} = \vec{x_j} - \vec{x_i}$. Initially, the ith robot is connected to its neighboring robots S_i through virtual springs, establishing the formation of the robots at the initial stage. Thus, this spring connection remains constant despite changes in distance between robots during the experiments. The alignment of the entire swarm determines the collective flock's performance. The degree of alignment, ψ, is the metric used to indicate the alignment status of the robots. A minimum value of $\psi = 0$ indicates the robots are non-aligned, while a maximum value of $\psi = 1$ means they are perfectly aligned.

Figure 4.2 shows two robots in the swarm and the resultant force $\vec{F_i}$ acting on one of the robots, determined from interactions with all neighbors. The figure indicates that the resultant force has a component perpendicular to the robot's current direction, causing rotation and eventually leading to alignment between the robots.

Although AES-based algorithms show promising results, virtual elastic interactions can cause fluctuations, leading to instability and reduced robustness, especially with noisy measurements. However viscoelastic interactions can mitigate these fluctuations, improving both stability and robustness. Thus, using viscoelastic links is expected to enhance flock motion while maintaining formation stability. In addition to the model's structure describing collective behavior, parameters play a crucial role in the swarm performance. These parameters are typically tuned empirically, but optimization techniques can significantly enhance performance. Therefore, several works have focused on optimizing AES model parameters to improve the performance of the computer model and make it more suitable for real-world applications using mobile robots. In the work of Raufi et al. [305], a

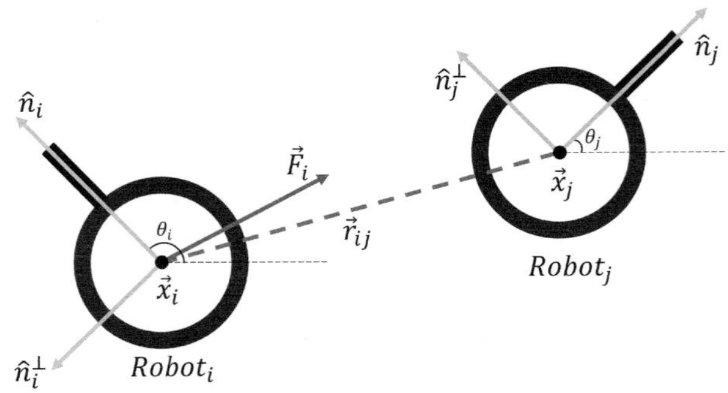

Figure 4.2: The schematic description of elastic interaction (from [26]).

Table 4.1: The values of the control parameters for AES, AES-TCACS, AES-GA and OCM algorithms.

	Model Parameters				
Methods	α	β	k	c	v_0
AES [116]	0.01	0.12	5	0	0.002
AES-TCACS [305]	0.066	0.97	1.28	0	0.05
AES-GA [27]	0.3696	0.9124	3.5329	0	0.05
OCM [28]	0.0262	0.5627	1.0	1.7503	0.075

Tabu Continuous Ant Colony System (TCACS) [193] was used to tune the AES model parameters by minimizing the force among robots and maximizing their alignment. The results showed better performance compared to the original parameters. Other studies [26, 27] improved the collective motion of the AES model using Particle Swarm Optimization (PSO) and Genetic algorithms, respectively. The cost function minimized virtual forces between swarm individuals, alignment error, and convergence time. This approach significantly enhanced collective motion behavior and swarm shape stability, outperforming both the original study [116] and the TCACS-optimized study [305]. However, these works did not consider measurement noise and its impact on collective motion behavior. In a recent study, Bahaidarah et al. [28] proposed an Optimized Collective Motion algorithm that employs viscoelastic interactions between robots to enhance robustness against various disturbances such as measurement noise, environmental factors, and modeling uncertainties. This underscores the algorithm's suitability for real-world robotic applications. The model parameters are automatically tuned using PSO optimization to achieve (i) minimal control effort, (ii) rapid alignment, and (iii) noise robustness.

Figure 4.3 illustrates four different experiments with swarms controlled by different algorithms listed in Table 4.1.

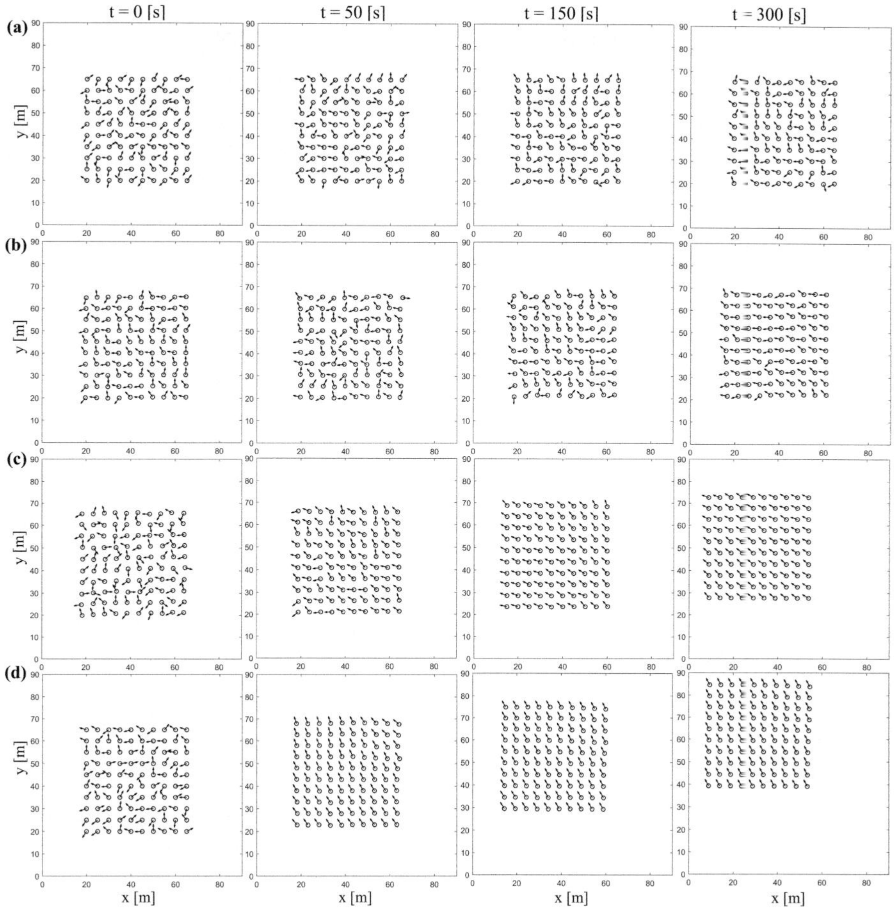

Figure 4.3: Swarm collective exploration controlled by (a) AES, (b) AES-TCACS, (c) AES-GA, and (d) OCM.

4.3 Use Case: Strategies to Mislead UAV Swarms

The core concept involves deploying defender UAVs into an attacking UAV swarm with the goal of misleading the swarm without causing further harm. Rather than relying on physical methods to remove the attacking UAVs from the sky or deploying defensive UAV swarms that engage in combat and risk collateral damage, we explore a strategy where a small number of defender UAVs infiltrate the attacking swarm to divert it from its intended mission. Figure 4.4 illustrates this scenario: (a) Shows a swarm targeting a specific objective, while (b) depicts defender UAVs infiltrating the swarm to mislead it. This approach has been largely overlooked in existing research [354], making

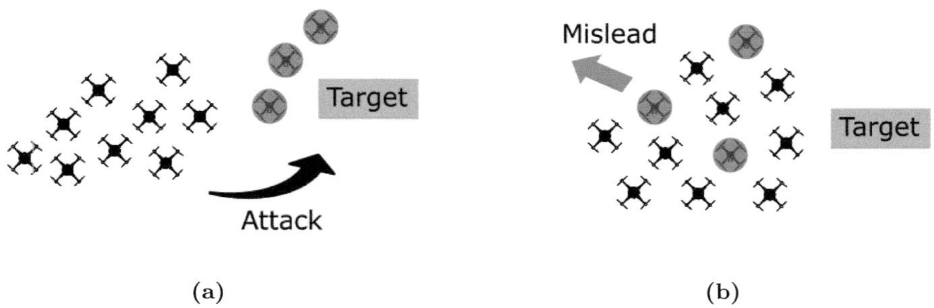

Figure 4.4: Main idea of the scenario, in which (a) UAVs are attacking a target and (b) defenders are induced into the attacking swarm to mislead it [354].

this work pioneering towards developing an intelligent and adaptable solution for redirecting swarm behavior intended for malicious activities.

In our initial exploration of how to divert a swarm from its mission, we equip both the attacking UAV swarm and the defending UAVs with the same algorithms. Our approach involves first *selecting* appropriate attacking algorithms and then *designing* effective defending algorithms. Given that the hostile swarm is assumed to be targeting a specific objective, we focused on target-oriented swarm algorithms inspired by nature, such as group hunting and foraging behaviors observed in animals and fungi. We selected the Grey Wolf Optimizer (GWO) [247] and the Slime Mould Algorithm (SMA) [228] for this study, both of which emulate natural strategies for locating food sources.

We equip the defending UAVs with the same algorithms as the attacking UAVs, slightly modified to divert the attackers from their original target. The primary goal of the defenders is to steer the attacking UAVs away from their target by influencing their movement. The scenario is set in a continuous 2D area containing a static target M at position $m = [x_m, y_m]$. Here, a swarm of attacking UAVs $a_i \in A$ (where $i = 1, \ldots, I$) and a smaller set of defending UAVs $d_j \in D$ (where $j = 1, \ldots, J$ with $J < I$) operate according to the local rules of their swarm algorithms. Each UAV is aware of its position, with $p_i = [x_i, y_i]$ for attackers and $p_j = [x_j, y_j]$ for defenders. The attacking swarm aims to locate and reach the target M using the Grey Wolf Optimizer (GWO) [247] and the Slime Mould Algorithm (SMA) [228]. All UAVs are equipped with processing, storage, and communication capabilities.

The defenders aim to redirect the attackers by modifying the fitness function q_m slightly from the original GWO and SMA functions. The modified fitness function also aims to minimize the distance to the target, with a smaller distance indicating better fitness. Fitness values are shared among all UAVs, allowing defenders to inject misleading fitness values to deceive the attacking swarm. The defenders must avoid injecting random values, as this would result in being disregarded by the attackers. Instead, they must move in conjunction with the attacking swarm, gradually diverting it from the target. The fitness

function for the defenders is modeled as follows:

$$q_m(p_j, m) = q(p_j, m) - \phi, \ \phi \in \mathbb{R}, \tag{4.3}$$

where

$$q(p_j, m) = \sum_{j}^{\dim p_j} (p_j - m)^2, \tag{4.4}$$

with p_j being the position of the defender UAV, m the position of the target, and ϕ a fixed value allowing defenders to achieve a better fitness relative to the attackers. The value of ϕ is arbitrary and determined through experimentation, where a low value results in minor distraction and a high value in significant distraction.

To effectively steer the attacking swarm, defenders need to move strategically. Preliminary experiments revealed that moving defenders to a fixed point, such as a static fake target, is ineffective due to the dynamic nature of the attacking swarm. Therefore, the next position $p_j(t+1)$ of a defender UAV is calculated as follows:

$$p_j(t+1) = p_j(t) + v_{\max}(t) \cdot (1 - c) \cdot \begin{bmatrix} \frac{1}{\sqrt{2}} \\ \frac{1}{\sqrt{2}} \end{bmatrix}, \tag{4.5}$$

where $p_j(t)$ is the current position, $v_{\max}$ is the maximum velocity of a UAV per time step, and c is the crowd factor, a value in the interval $[0, 1]$. The crowd factor balances between leaving the attacking swarm ($c = 0$) and following it ($c = 1$), allowing defenders to stay close enough to influence the attackers gradually. The vector $[\frac{1}{\sqrt{2}}, \frac{1}{\sqrt{2}}]$ is of unit length but can be oriented to direct UAVs towards a specific direction (e.g., bottom right).

Figure 4.5 presents the results of the GWO simulations, illustrating (a) the probability h_P of attackers winning as the number of defenders increases, and (b) the discrete probability density of the achieved hits over time with varying numbers of defenders. As anticipated, the presence of more defenders improves h_P, as a greater number of defending fake-wolves exerts a stronger influence on the entire swarm. Details of the results can be found in a paper by Simonjan et al. [354].

4.4 Use Case: The Principle of Hormones for Production Plants

As already described in Section 3.2, the production of logic and power integrated circuits (ICs) in the semiconductor industry is inherently dynamic and complex [138]. Unlike the high-volume production of memory ICs, the logic and power sector features wafer production with a vast product mix, frequent system changes, and a large number of processing steps involving numerous machines [203]. Weekly operations can involve approximately

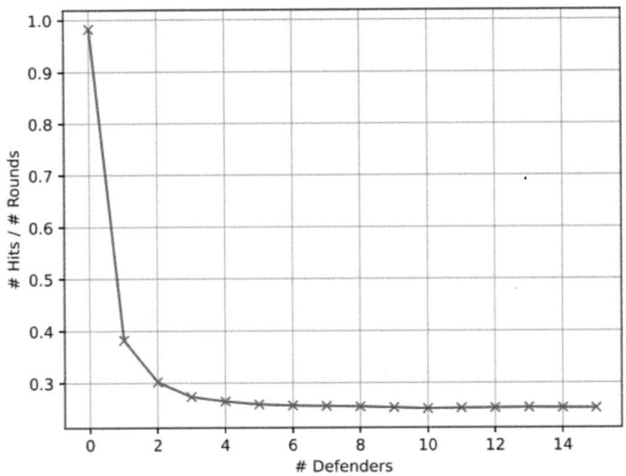

(a)

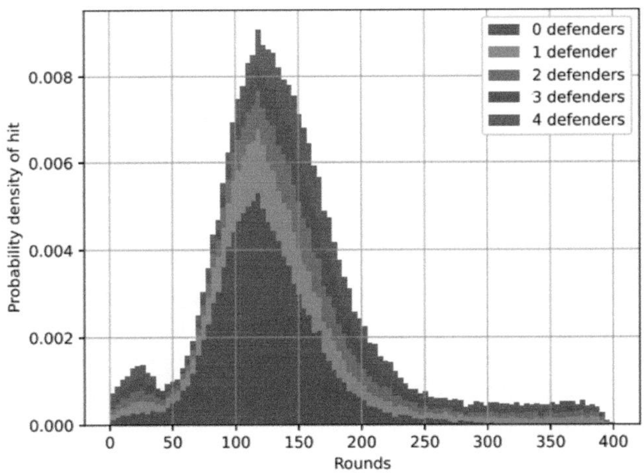

(b)

Figure 4.5: GWO simulation results: (a) Probability h_P that attackers win over an increasing number of defenders and (b) probability density of achieved hits over time and increasing number of defenders [354].

10^5 tasks across 10^3 machines [379]. Optimizing this process for work in progress (WIP) and flow factor presents an NP-hard challenge [130]. Existing dispatching rules and linear optimization methods struggle to handle the NP-hard search space [222], limiting their ability to account for the entire system behavior due to computational complexity, and thus failing to fully exploit optimization potential in job-shop scheduling within semiconductor production systems [203].

To address these challenges, we propose modeling the production plant as a self-organizing system of agents that interact non-linearly, as described in Section 3.2. This approach aims to achieve near-optimal solutions within feasible computation times. Such a system can adapt to environmental changes, scale with the number of agents, and maintain robustness against single points of failure due to its reliance on local interactions [160]. Utilizing local rules and interactions helps circumvent the extensive computation times associated with centrally performed linear optimization.

One promising self-organizing approach involves algorithms inspired by biological hormone systems. These artificial hormone systems mimic the biological endocrine system, which regulates cellular metabolism in the body [358, 387]. As part of the broader class of self-organizing systems, they exhibit properties like scalability, adaptability, and robustness [292], making them suitable for coordinating complex agents in networked technical applications [43, 98, 356]. Such algorithms are particularly valuable for large cyber-physical systems, such as those found in the semiconductor industry, where traditional control or scheduling mechanisms reach their operational limits.

The artificial hormone algorithm is designed using a bottom-up approach to represent both the urgency of processing a lot and the demand for incoming lots by machines. In this system, artificial hormones are generated at machines and diffuse through the production environment, aligning with the processing steps of the lots. The lots function as swarm members that are attracted to machines based on the hormone levels present. The algorithm relies on five key parameters, each with specific mechanisms described below. Table 4.2 provides an initial set of parameter values used for preliminary simulations. The algorithm outlines the calculations and decision-making processes that influence the handling of lots, offering flexibility in implementation. For instance, hormone-related computations can be performed directly at the machines or within a networked monitoring and control layer [101].

Hormone Model Each processing step is associated with a specific hormone type. Hormones can exist at any machine, and multiple hormone types can coexist at a single machine. At each simulation tick, hormones degrade exponentially at a rate α:

$$hormone\_amount = hormone\_amount \cdot (1 - \alpha) \tag{4.6}$$

An evaporation rate of 0 means that hormones do not degrade, while the maximum value of 1 indicates immediate degradation.

Hormone Production by Machines to Attract Lots Each machine produces a hormone corresponding to its process type. Machines performing the same process type produce the same hormone. If a machine handles multiple process types, it generates hormones for each type in equal parts.

Table 4.2: Suggested algorithm parameters.

Parameter	Value
α	0.3
β	1
γ	0.5
δ	0.2
ε	0.8

Machines aim to maximize their working time by attracting enough lots into their queues. The hormone output of a machine is calculated as follows:

$$hormone\_output = \frac{1}{lots\_in\_queue + \beta}, \tag{4.7}$$

where β is a smoothing factor greater than zero.

Machine Linking A machine A is upstream-linked to machine B if a recipe includes processes of both machines in consecutive steps. If each machine performs only one process, the link strength from this recipe is 1. Otherwise, the link strength is 2 divided by the number of processes supported. If the sequence appears in other recipes, the link strength accumulates.

Hormone Diffusion Upstream When upstream links exist, a fraction γ of the hormone at a machine diffuses upstream:

$$upstream\_hormone = hormone\_amount \cdot \gamma \tag{4.8}$$

$$hormone\_amount = hormone\_amount - upstream\_hormone \tag{4.9}$$

Each upstream-connected machine receives a proportional share of the upstream hormone

$$added\_hormone = upstream\_hormone \cdot \frac{link\_strength}{\sum link\_strengths}, \tag{4.10}$$

where *link_strength* refers to the upstream link strength for the given hormone between the respective machines, and $\sum$ *link_strengths* denotes the sum of all upstream link strengths for the hormone from the sending machine.

Hormone Diffusion by Incoming Lots Incoming lots cause a portion δ of the corresponding hormone at a machine to diffuse upstream:

$$upstream\_hormone = hormone\_amount \cdot \delta \tag{4.11}$$

$$hormone\_amount = hormone\_amount - upstream\_hormone \tag{4.12}$$

The upstream hormone is added to the machine from which the lot arrived, allowing the flow of lots to self-stabilize:

$$added\_hormone = upstream\_hormone \tag{4.13}$$

Lot Prioritization by Timing A lot's base priority is determined by its remaining Raw Process Time (RPT) and Planned Cycle Time (PCT):

$$base\_priority = \frac{remaining\_RPT}{remaining\_PCT} \tag{4.14}$$

Hormone Attraction of Lots The priority of a lot is influenced by hormones present at the machine where the lot is waiting:

$$attraction = \sum_{i=0} h_i \cdot \varepsilon^i, \tag{4.15}$$

where h_i represents the hormone of the process that is i steps ahead in the lot's recipe, with h_0 being the hormone of the current process. The factor ε indicates the influence strength of the hormone.

The lot's priority is then calculated as:

$$priority = base\_priority \cdot attraction. \tag{4.16}$$

At each machine, lots are processed based on their priority, and if batch processing is used, all eligible lots are processed together in a batch.

The algorithm was implemented and tested using a NetLogo simulation model. The simulation results show that the artificial hormone system improves overall production time and the flow factor by approximately 5%. Further details and the evaluation of the algorithm can be found in Elmenreich et al. [101].

4.5 Design Behaviors using the Concept of Evolution

When designing a swarm robotic system, several decisions must be made, including the choice of the robot model and its capabilities, the planned number of robots, the control system, the software model, and the algorithms to be employed. A major challenge is given by the fact that decisions on one aspect can influence the possible degrees of freedom at another level. For instance, choosing a simple robot with a limited number of sensors would necessitate a control system that relies on the close cooperation of a large number of robots to accomplish the task. The challenge of determining the system architecture at multiple levels is akin to the hardware-software co-design approach used for embedded systems [407].

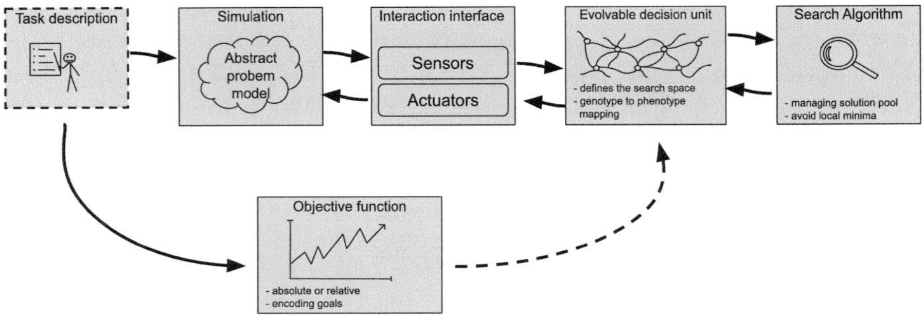

Figure 4.6: Proposed design methodology [112].

Following a design process supported by evolutionary optimization, we propose a methodology that addresses the key decision points faced by designers of robot swarm experiments. For a given experiment, we identify five major components to be designed, as depicted in Figure 4.6. These five major components are derived from the system architecture of a self-organizing system design process described in Fehervari and Elmenreich [112]:

Task description: Set of requirements that the solution has to meet.

Simulation setup: Describes the simulation model and the relations between the system's components and their environment.

Evolvable decision unit: The evolvable representation of the local rules.

Interaction interface: Describes how the decision unit interacts with its environment.

Search algorithm: A meta-heuristic algorithm responsible for finding novel and better solutions.

Objective function: The cost function guiding the search algorithm.

Engineering problems usually begin with a task description clearly outlining the goals and constraints. This description serves as a contract outlining the expectations of the desired system at a high level of abstraction. Thus, the task description significantly influences the simulation model used and the target function. In a further step, an efficient system model is developed based on the task description. This model should accurately capture the essential aspects of the system to be simulated while abstracting from unnecessary factors. It is important to note, however, that this step should not delve into how individual components are rendered or how they interact, as that would preempt some important decisions. For this reason, we divide the process into two units: The interaction interface unit and the

evolvable decision unit. The former focuses on strategically planning the interactions between system components and their environment, specifying communication methods (such as sensors) and the underlying interfaces (such as protocols). The latter focuses on the actual representation of the system components, which may include one or more model types depending on the homogeneity of the system. The reason for separating this part from the system model is the need for an evolutionary approach, which requires "evolvable" representations. In other words, components must be able to generate adaptive genetic diversity, which is typically achieved by defining mutation or crossover operators [10]. Given the variety of available representations, careful consideration by the designer is essential because of the often different properties associated with each representation.

An important part of the evolutionary process is the algorithm that optimizes candidate solutions. This can be viewed as an efficient search algorithm that operates on candidate solutions represented by the component representation. Both the component representation and the search algorithm can be implemented independently against a well-defined interface defining how the search algorithm can access and evaluate candidate solutions. This allows optimization and candidate representation to be chosen separately. There can be an advantage of using a particular search algorithm with a particular candidate representation; for example, evolving neural network structures would benefit from a niche formation mechanism that prevents innovations from dying out prematurely, and restricting recombination to compatible descendants among the pool of neural network solutions [363].

To guide the evolutionary process towards a desired solution, an objective (or fitness) function needs to be defined that either indicates the quality of a particular candidate solution or at least allows for the comparison of the quality of two candidate solutions. Designing a proper objective function is crucial for the overall evolutionary optimization approach. The objective function should be readily accessible, either continuous or sufficiently fine-grained, and should clearly favor the desired solutions while avoiding the reward of undesirable ones. A good objective function should not only assess high scores to proper solutions but also form a smooth fitness landscape, avoiding, whenever possible, local maxima that could trap the evolutionary search. Designing such a practical function is often difficult, even for domain experts [183].

4.6 Framework for Evolutionary Design

4.6.1 Architecture

The architecture of FREVO is strictly component-based, where the steps of evolutionary design are divided into individual components. This architecture makes it possible to develop a single component and easily replace individual components. Therefore, different configurations can be easily evaluated to find the most appropriate configuration for a given problem. Each component

implements a particular feature of the evolutionary approach. The *problem* component defines the specifics of a CPS controller, environment, and fitness function. The *representation* component defines how the CPS controller is represented. The *optimization* component defines the method for finding the optimal candidate representation. The *ranking* component defines how the candidate representations are ranked based on their performance.

FREVO is implemented in the Java programming language and follows the object-oriented programming paradigm for encapsulating components. Each component is defined by an abstract class, standardizing the interfaces between parts such as for example agent representation, simulation and optimization algorithm. Therefore, new components can be used in a setting with existing components completing the overall system. Creating a component is guided by a built-in component generator, which helps the software developer generate the required code framework in the context of the class hierarchy. FREVO is released under an open source license, supporting the sharing of research ideas and technical solutions. The source code is freely available.[1]

The FREVO GUI guides a user step by step through the configuration process. This is done by selecting a component for each task of the evolution process. A single configuration of components is called a FREVO session. Such sessions and the results of the optimization process can be exported and saved for import for later use.

Problem Definition

The problem definition in FREVO defines the parts related to the task to be achieved, including the environment and the interaction possibilities of the agents, but not the implementation or optimization of the behaviors. The problem definition thus consists of the formulation of a goal, the simulation model, and the interaction interface for the agents within the simulation. In the case of CPS systems, the interaction interface models the behavior of sensors and actuators, and the environment is implemented as a physical simulation model. Candidate controller representations are linked to sensors and actuators, with each sensor serving as input and each actuator as output.

The problem definition also implements an objective function and the means for evaluating the values of the objective function in the course of one or multiple simulation runs. Thus, the fitness assessment typically occurs in the phenotype space, where the behavior in the simulated environment is observed. The genotype, which refers to the implementation of a controller, typically cannot be assessed without executing it in a simulation. The objective function then guides the heuristic optimization process in finding the optimal candidate representation.

There are two problem definition types: *AbstractSingleProblem* evolves a CPS candidate for cooperative task execution in a swarm setup,

[1] FREVO Github: https://github.com/smartgrids-aau/Frevo [Online; accessed: 05-February-2025].

while *AbstractMultiProblem* involves evolving and evaluating multiple representations against each other, suitable for competitive multi-agent systems. The former applies to homogeneous multi-agent systems, ranking candidates by fitness, while the latter suits competitive systems, with fitness evaluated relative to other agents. Tournament algorithms rank candidate pools, as seen in scenarios like soccer games: To determine the best team among several soccer teams, the teams play each other until a tournament winner is determined. An example of a soccer team use case, where two teams compete against each other in an evaluation, is presented in [111].

Developing a new problem definition involves defining the interface between sensor inputs, actuator outputs, and candidate representation; implementing simulation for evaluation; and calculating fitness based on performance measures, all guided by predefined system interfaces. Given existing components, developers can focus on implementing new problems without concern for representation, optimization, or ranking components. Complex simulations can leverage external simulators from model libraries or custom interfaces as detailed in Section 4.6.4.

Candidate Representation

The candidate representation models the internal structure of the CPS controller. It is a generic structure that is evolvable, for instance, an Artificial Neural Network (ANN). It encodes the behavior of the CPSs, including reactive behavior to stimuli via the sensors. A candidate representation, when viewed together with the problem definition, represents a possible solution to the given problem. Every representation must define the genetic operators such as mutation, crossover and selection. For supporting the user in analyzing the representation, different output formats can be implemented. Typically, the representation is derived from the *AbstractRepresentation* class and is common among all agents. For heterogeneous multi-agent problems, one can choose the bulk representation to evaluate a set of candidates with distinct representations.

Currently, FREVO supports the following representations:

- *Fully-meshed net*: This is a recurrent ANN with a fully meshed architecture providing mutual connections between any two neurons. Throughout evolution, both the biases of the neurons and the connection weights undergo alterations. During run-time, this network has some memory due to possibilities of feedback loops.

- *Three-layered net*: This is a feed-forward, non-recurrent ANN incorporating one hidden layer. The biases of the neurons and the connection weights evolve similarly. Unlike the fully meshed ANN, this architecture is tailored for simpler problems, substantially reducing the search space.

- *NEAT*: In this model, the ANN adapts by evolving the connectivity between neurons, following the NeuroEvolution of Augmenting Topologies (NEAT) method introduced in [363].
- *HebNet*: Incorporating Hebbian learning, this recurrent, fully interconnected ANN endows synapses with plasticity enabling real-time learning. In this setup, both plasticity and initial weights are evolved with Hebbian learning occuring during the simulation runs.
- *Simple bulk representation*: This is a combination of multiple representations as described above.

Optimization Method

The optimization method aims to find the candidate representation that maximizes fitness, as the problem definition specifies. It employs genetic operators from the candidate representation to generate new candidates in each generation, replacing the least effective candidates. This iterative heuristic search progressively produces candidates with improved performance. The search continues until a termination criterion specified within the optimization method is met. Examples of termination criteria include reaching a maximum number of generations or consecutive generations without fitness improvement.

The optimization methods currently offered by FREVO are as follows:

- *Random search*: A baseline comparison technique in which randomly generated candidates replace those with low fitness.
- *NNGA*: This is an Evolutionary Algorithm (EA) that maximizes the population diversity. It supports multiple populations and several ranking algorithms. It is based on the Neural Network Genetic Algorithm (NNGA) described in [100]. It is well-suited for evolving any type of representation.
- *GASpecies*: This is an EA that classifies candidates into species. Species are determined by a similarity function defined in the candidate representation. Within each species, candidates share the same fitness value.
- *CEA2D*: It is a cellular EA that arranges all candidates on a 2D torus surface. Genetic operations are executed in a local context, resulting in improved diversity and slower convergence compared to traditional evolutionary algorithms.
- *Novelty search*: This is an EA that prioritizes behavioral diversity over fitness. Its implementation is based on rtNEAT in [362].
- *Novelty species*: This is an EA that rewards behavioral diversity across different species.

Ranking Algorithm

The ranking algorithm sorts candidate representations based on their performance, i.e., fitness value. The ranking algorithm is also responsible for parallelization to decrease the overall simulation time of the optimization process by deciding which evaluations can run in parallel on a multi-threaded system. Problems where the fitness of a candidate can be obtained directly by a simulation run require a so-called *absolute ranking*. Two types of absolute rankings are currently implemented in FREVO:

- *Absolute ranking*: A ranking algorithm that sorts candidates by the fitness value returned from the problem component. It supports multi-threading to decrease the time needed for optimization.
- *Novelty ranking*: A ranking algorithm that sorts candidates based on their novelty in the behavioral space.

In contrast, problems that necessitate pairwise evaluations between the candidates follow the structure of an *AbstractMultiProblem*. For this type, a full tournament ranking and a ranking based on the Swiss system are provided. The Swiss system, inspired from chess tournaments, can provide a ranking with fewer comparisons at the cost of ranking accuracy [99].

4.6.2 Graphical User Interface

The GUI of FREVO provides a convenient means to assess an evolutionary design process component swiftly, showcasing the underlying modular architecture (see Figure 4.7). This visual representation illustrates FREVO's GUI, where an example problem has been evolved, with the "Select Problem Component" window displayed.

The top-left "Configure Session" panel facilitates session configuration by guiding users through selecting a problem, optimization algorithm, candidate representation, and ranking method step by step. This streamlined process expedites the testing of new components. Each selection prompts a new window for configuration, with options for fine-tuning parameters.

Below, the "Control" panel enables users to initiate, halt, or reset the optimization process. Progress can be tracked in real-time through the "Statistics" panel on the right and the "Console" panel below. The "Statistics" panel presents graphs illustrating the evolution of fitness and diversity across generations, while the "Console" panel displays output from active components. Upon completion of optimization, results can be accessed from the panel on the left that let's the user inspect the latest generation. Here, users can save or replay the results for further analysis. Clicking "Replay" allows a closer examination of each candidate representation from the last generation, including visualization of resulting behavior in simulation.

The top menu provides options for saving and restoring previous sessions and managing components, including the "Component Creator".

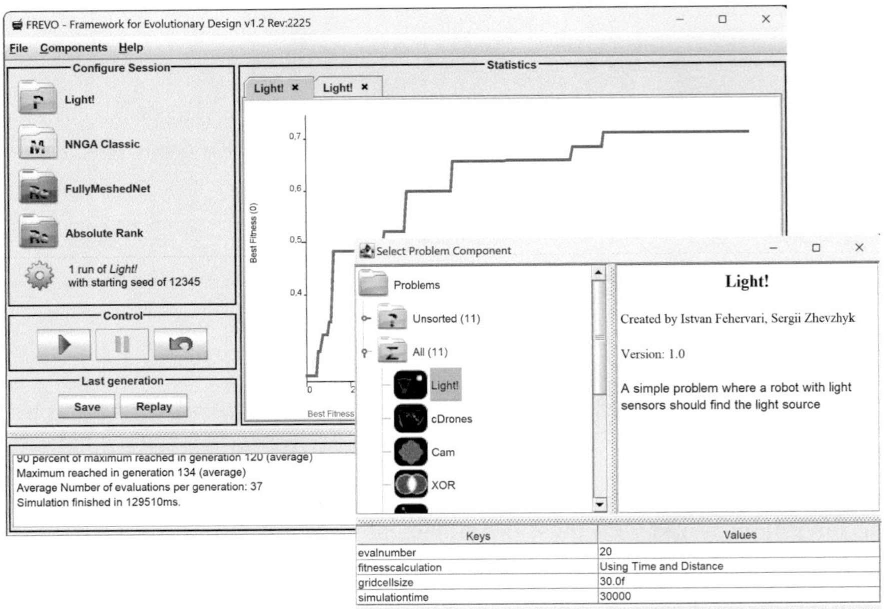

Figure 4.7: Screenshot of the FREVO GUI showing the evolution of an example problem.

This tool automates skeleton code generation, aiding software developers in implementing new components for the model library.

4.6.3 Workflow

When executing FREVO with existing components, initiating optimization requires just a few clicks. First, users select the desired problem, followed by the optimization method, candidate representation, and ranking algorithm. Optimization continues after clicking the play button until the selected termination criterion is met. The statistics panel displays two graphs: One depicting the best fitness value of each generation and another showing the diversity in each generation (refer to Figure 4.7).

The component creator in the top menu can be utilized to model a new problem. A code skeleton is generated after selecting the component type, name, package, and description. This code is placed in a subdirectory of the components directory in FREVO, accompanied by comments guiding further implementation (see Figure 4.8). Additionally, an XML file named after the component is created, allowing the definition of configuration parameters, sensor inputs, and actuator outputs for the CPS.

The primary task involves implementing the *evaluateCandidate* method, where the candidate representation is evaluated through simulation, either

```
*MyProblem.java
package myproblem;

import core.AbstractSingleProblem;

public class MyProblem extends AbstractSingleProblem {

    @Override
    /** Evaluates the given representation by calculating its corresponding fitness value. A higher
     * @param candidate The candidate solution to be evaluated.
     * @return the corresponding fitness value. */
    public double evaluateCandidate(AbstractRepresentation candidate) {
        // TODO Auto-generated method stub
        return 0;
    }

    /** Returns the achievable maximum fitness of this problem. A representations with this fitness
    public double getMaximumFitness() {
        return Double.MAX_VALUE;
    }

}
```

Figure 4.8: Screenshot of the source code skeleton of a newly created problem.

directly within FREVO or by invoking an external simulator. Consequently, the environment and CPS must be implemented. Sensor inputs of the CPS are passed to the *getOutput* method of the candidate representation, which returns the actuator output(s). A suitable performance measure is then implemented to compute the fitness value of the simulation run.

Once a component's implementation is complete, it needs to be compiled and automatically loaded upon FREVO launch. The new component then appears in the component selection window, enabling the aforementioned workflow. This process remains consistent when creating other components, with the generated code skeleton and implementation tailored to their specific requirements.

4.6.4 External Simulators

External simulators are used in domain-specific simulations. To achieve this, the *evaluateCandidate* function of the problem component must invoke the simulator and provide the candidate representation (i.e., the CPS controller). Consequently, the simulator must incorporate the exact representation to enable FREVO to evolve an optimal solution. This can be accomplished through two approaches: In a compact way through code generation that is then run in the simulation, or in a modular fashion, by transmitting only the inputs and outputs between FREVO and the simulation.

In the compact approach, the simulator's code needs to be recompiled with the newly created representation in each generation. Subsequently, the simulator is executed directly from FREVO, returning the fitness value for optimization. This method is preferable for CPSs lacking file system or network communication capabilities.

Alternatively, the modular approach entails implementing the simulation after achieving the desired candidate representation. In each generation, the parameters defining the representation are then transmitted to the simulator, either through files or network communication. The actions performed by

individual CPSs within the simulator are fully determined by the candidate representation, whose parameters are evolved by FREVO. FREVO executes the simulator, which logs the performance measure into a log file. Upon completion of the simulation, the corresponding fitness value is computed by the *evaluateCandidate* method.

4.7 Why Evolutionary Optimization Needs Simulation

In this chapter, we have explored the design of self-organizing systems through evolutionary methods, focusing specifically on swarm robotics. We introduce the FREVO framework, which leverages evolutionary optimization to address challenges in decision-making and hardware-software co-design. FREVO is a modular component-based system that can be adapted for various applications. Its components are categorized to facilitate simulation, representation of candidates, optimization methods, and algorithm ranking. The system includes a user-friendly GUI for configuring sessions, selecting components, and executing the optimization processes.

Typically, evolutionary approaches rely on assessing the value of a fitness or cost function for a given implementation. This often involves conducting experiments where swarm members interact within an environment similar to their intended operational context, collaborating to solve typical problems. However, fitness values typically are obtained by post-experiment analysis, which in turn demands reinforcement learning techniques capable of handling delayed rewards. While theoretically feasible with real hardware, the practical constraints of conducting thousands of experiments can be prohibitively cumbersome. The challenges include not only the logistical issues of experimentation but also the substantial costs associated with deploying a large number of capable swarm robots, especially when such robots are not mass-produced. Consequently, simulation becomes an indispensable tool in swarm engineering.

In the upcoming chapter, we will explore the pivotal role of simulation in the development and analysis of swarm robotics and cyber-physical systems. We will examine why simulation is an essential tool for advancing our understanding of swarm dynamics and overcoming the practical challenges associated with real-world testing, which can be complex, costly, and logistically demanding. The chapter will provide a comprehensive overview of different swarm simulation techniques, including agent-based models and multi-agent systems. We will review a range of simulation tools and platforms, focusing on their capabilities, advantages, and limitations. Emphasis will be placed on the practical aspects of using these tools, such as cost efficiency and the ability to handle large-scale swarm scenarios in a time-efficient manner, making them especially valuable for researchers with limited resources. We will also address the inherent challenges of swarm simulation, including

the limitations in modeling accuracy and the difficulties in replicating real-world conditions. The discussion will cover how simulation integrates with hardware development to validate results and how emerging technologies, like AI and machine learning, are transforming the landscape of swarm simulation. Through this exploration, we aim to shed light on how simulation can effectively complement physical experiments, offering insights and solutions for designing, testing, and optimizing swarm systems. We will highlight both the benefits and drawbacks of simulation platforms, drawing on examples such as the bio-inspired swarm aggregation mechanism BEECLUST and its various simulations across different frameworks. By understanding these dynamics, researchers can better navigate the complexities of swarm robotics and cyber-physical systems, ultimately advancing the field with more innovative and cost-effective approaches.

Chapter 5

Simulating Swarms of Cyber-Physical Systems

Real-world implementation and evaluation of a swarm robotics scenario that often requires a huge number of robots is very complex, costly, and requires large settings. Therefore, simulation platforms for swarm systems are a very popular and efficient choice for a research project in an early stage. This motivated researchers to develop many swarm simulation platforms that successfully replicated the bio-inspired mechanisms of swarm robotics. These simulation frameworks, predominantly open-source, have enabled several research groups, even those with limited budgets, to conduct swarm studies involving a large number of robots.

There are many benefits and drawbacks in using simulation software instead of a real-world robotic platform. The main advantage is the cost efficiency of the simulation platforms that are mostly designed for open-source operating systems (OS) such as Linux. Although their speed and performance are limited to the host system performance, they can implement complex swarm tasks in a low-cost, time-efficient manner. Additionally, simulation software offers various benefits that make them an ideal choice for early career researchers and students, such as the ability to share code and experimental configurations with other researchers. This process allows to easily maximize the replicability of research works. The biggest drawback in using simulation software is the poor modeling and reduced abstraction that introduces inaccuracy when simulating large numbers of robots in a swarm scenario. Even internal and external conditions are hard to model and often present a poor environment for virtual swarm robotics platforms. Additionally, there are many uncertainties and complexities at various levels, from individual behavior to collective behavior. Consequently, the results obtained from simulation software often do not align with the outcomes observed in real-world-robot experiments.

Table 5.1: List of simulation platforms commonly used for swarm robotics.

Simulation Platform	Operating System (OS)	Level	Open-source	Application	2D/3D
AirSim [348]	Linux/Win	Physics	Yes	Generic	3D
ARGoS [290]	Linux/Mac	Physics	Yes	Swarm	2D & 3D
BeeGround [229]	Linux/Mac/Win	Physics	Yes	Swarm	2D & 3D
Gazebo [209]	Linux/Mac/Win	Physics	Yes	Generic	3D
Kilombo [179]	Linux	Abstract	Yes	Swarm	2D
MASON [236]	Linux/Mac/Win	Abstract	Yes	Generic	2D & 3D
MESA [239]	Linux/Win	Abstract	Yes	Generic	2D
NetLogo [382]	Linux/Win	Abstract	Yes	Generic	2D
OpenHRP [190]	Linux/Win	Physics	Yes	Generic	3D
PyCX [324]	Linux/Win	Abstract	Yes	Generic	2D
SCRIMMAGE [81]	Linux/Mac	Physics	Yes	Generic	3D
Stage [392]	Linux/Mac	Physics	Yes	Swarm	2D
Swarm-sim [60]	Linux/Mac/Win	Abstract	Yes	Swarm	2D & 3D
USARSim [56]	Linux/Mac/Win	Physics	Yes	Generic	2D & 3D
V-rep [315]	Linux/Mac/Win	Physics	No	Generic	3D
Webots [246]	Linux/Mac/Win	Physics	Yes	Generic	3D

As an example for the extensive work on the implementation of a robotic swarm scenario, the bio-inspired swarm aggregation mechanism, BEECLUST [327], has been simulated numerous times with different settings in various simulation frameworks. These setting include pheromone-based aggregation in PyCX simulator [23], a study of swarm interaction in NetLogo software [37], fuzzy-based decision making in Stage [22], aggregation in a complex environment in BeeGround [401], and a source exploration scenario in Webots [11]. However, while these studies implemented the same bio-inspired aggregation behavior, they did not consider the honeybees' diversity (behavioral and physical heterogeneities). Hence they did not replicate similar results observed from real-life honeybee and real-robot swarm experiments [374, 329] as this abstraction level was already too high. Similar limitations were also seen in other swarm algorithms, such as flocking.

Table 5.1 lists some of the popular softwares that have been created for the usage of simulating CPS swarm scenarios and applications. Simulation softwares can model agents at various complexity levels, ranging from abstract simulators like PyCX and NetLogo that do not represent the physical properties of real robots, to physics-based realistic simulators like ARGoS and Gazebo which consider all the physical parameters of a mobile robot. The reality level of the simulator directly relies on the PC's processing power. Nowadays, high-performance PCs allow us to implement large swarm experiments with physics-based simulation very close to reality.

5.1 Simulation Requirements

The main reason for using simulation software instead of real robots is to have a cost-effective approach to investigate new ideas in a time-efficient manner. While a trade-off exists between the precision of implementation and associated costs, the consensus among researchers is that in the initial phases

of a project involving swarm scenarios, simulations represent the optimal approach. As a result, simulation software for CPS swarms are becoming increasingly popular among small and large research labs. Therefore, to perform a reliable swarm experiment in simulation, the utilized simulation platform must have the following criteria:

1. It must support a fully **decentralized** process in which each robot can have its own control system and parameters locally. It is very important that each robot in the simulation platform operates in a decentralized manner with its own internal controller. Although having a central controller can simplify the implementation of a swarm experiment, replicating this setup in a real-world scenario would be complex especially when dealing with a large number of robots or when robots are deployed over vast distances such as 100s of kilometers apart.

2. It should consider the **physical properties** of a real-world robotic platform as much as possible, including aspects such as motors and gears, friction, mass, power consumption, and more. Nonetheless, it is crucial to bear in mind that the ultimate objective is to deploy swarm behaviors in real-world scenarios, resolving human problems effectively in the practical application phase. Therefore, a realistic simulation of a swarm system will allow us to replicate the behaviors of the swarm on real robots with minimum effort.

3. It should replicate the physical properties of the real-world **environment**, such as light, shadow, color, and texture. Similarly to the previous case a simulation must be capable of handling environmental conditions and replicating physical properties as closely as possible to real-world conditions.

4. The simulated **sensors** must be realistic enough to model various physical measurements such as velocity, force, position in 2D and 3D, and heading angle. However, the real-world implementations always include some level of false data due to noise in the sensory system. Hence, it is instructive to consider this noise incorporating both systematic and random errors.

5. It should support **long-term autonomy** and provide enough memory and processing power so that a swarm can converge to a stable state. This is also important when we implement the swarm behavior using real robots that need to work hours and hours before recharging their batteries or transferring recorded data to an external storage facility.

6. Last but not the least, the simulator must support **open-source** development so that researchers can customize it to fit their applications. Commercialized simulation software that lacks support for open-source development will not serve as a versatile platform for everyone, from early career researchers to established researchers.

A CPS swarm simulator must follow these essential criteria to enable a reliable swarm experiment and be an attractive simulation platform that maximizes replicability. However, the significance of each criterion depends on the type of simulation output being planned. For example, in using a simulator to act as a 3D space containing particles like robots acting as gas molecules, the physical consideration is limited to molecules' properties rather than defining motors and gears. Therefore, we first need to define whether we go for an abstract or a physics-based simulation software.

5.2 Abstract Simulation

Abstract simulation tools, also known as kinematic-based or particle-based simulations, are powerful tools for the conceptual investigation of swarm behaviors. However, these simulation platforms do not consider the physical properties of the systems; hence, they allocate entire processing power on executing inter-agent interactions in a very large-size swarm. One of the early swarm simulation examples was Boids model developed by Reynolds [311]. The swarm flocking was achieved through three behaviors I.e., collision avoidance, velocity matching, and flock centering. These behaviors were implemented with the assumption that individual agents have access to the distances and directions of their local neighbors. Other related studies proposed mathematical models describing the collective motion of the swarm, such as the model proposed by Vicsek et al. [394]. In a very interesting work by Schmickl et al. [328], they used an abstract simulation to show a fascinating behavior of a life-like system. They simulated thousands of particles in tens of thousands of simulation runs to show emerging patterns of artificial cell life cycles. In this kind of simulation, each agent is a particle in a 2D or 3D space that freely roams and interacts with others. Parameters such as linear and angular velocities, radius of sensing, size of the agent, basic dynamic model, and more, are yet to be considered in the experiments. In another study, Ferrante et al. [116] introduced a self-propelled collective motion mechanism using a two-dimensional active elastic sheet (AES) model to simulate swarm agents as particles using an abstract simulation. This type of particle-based setting was also used to model interaction networks for collective motion [386]. The AES model was adopted in other studies and several extensions of the model with optimization approaches were implemented using abstract simulations [28, 26].

Abstract simulation is extensively utilized to explore control engineering mechanisms implemented at a theoretical level. These studies are typically conducted using general programming platforms such as MATLAB®. For example, in [172], a swarm coordination protocol was investigated considering the non-linearity of the agents' dynamics, which mainly focused on the design of a robust control system for robotic swarms. In another study focused on the robotic shepherding task [171], an abstract simulation of sheep and dogs was developed considering the real dynamics of shepherding dogs to the sheep. The

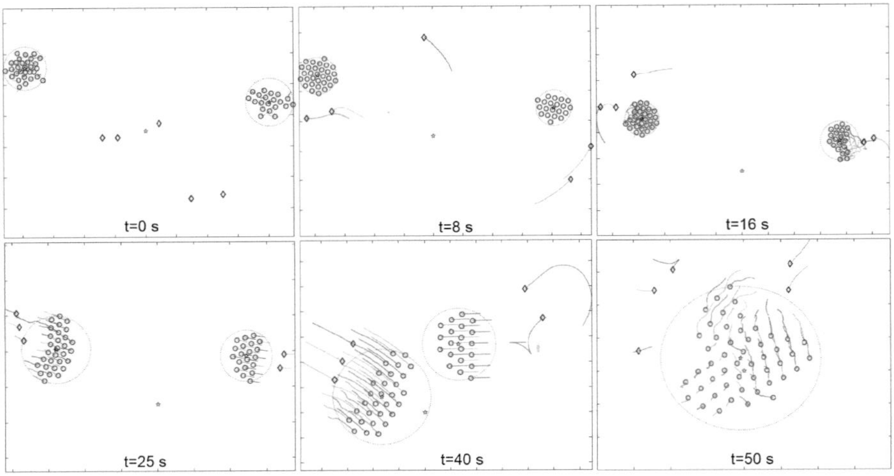

Figure 5.1: Example of a shepherding scenario in an abstract simulation. The positions of the dogs are shown with blue diamonds and the sheep with red circles. The target point is marked by a red star. At the beginning, the dogs were scattered around the target area. From 25 to 40 seconds, the sheep were guided towards the goal. After 40 seconds, one group of sheep had arrived and the dogs began to assist the other flock. At 50 seconds, the two flocks of sheep were combined into one large herd and the herding task was completed.

simulation output offered valuable insights before the actual implementation of the robotic system. Figure 5.1 illustrates the shepherding control strategy and the agents' trajectories in an abstract simulation. Similar implementation of the theoretical swarm system was also utilized in several research works including, formation control [409, 408, 410].

Many studies opt for abstract simulation due to the ease of implementing swarm systems in this environment. In this chapter, we introduce an open-source abstract simulation framework programmed in NetLogo, SwarmFabSim [390].

5.2.1 SwarmFabSim: A NetLogo Implementation

NetLogo is a popular agent-based simulation platform widely utilized for simulation of self-organized systems in research and education [404]. It is a freely available tool that boasts extensive documentation and is actively maintained, ensuring a robust and stable code base. NetLogo provides users with numerous extensions, further expanding its functionality and versatility for various simulation purposes. The simulation software is widely recognized for its extensive use in educational settings, particularly for teaching agent-based modeling and complex systems. However, it is important to note that NetLogo is not limited to educational purposes only. It has demonstrated its maturity as a platform capable of conducting simulations involving several

thousand agents in a reasonable computing time. This has been confirmed through studies such as those by Railsback et al. [303] and [302], which emphasize the platform's ability to handle large-scale simulations effectively. For our own performance results we refer the reader to Umlauft et al. [390]. The NetLogo homepage proudly showcases a vast collection of over 3000 research papers published in the last decade that have employed NetLogo as their simulation platform of choice. This extensive list is a testament to the widespread adoption and credibility of NetLogo within the scientific community for conducting a diverse range of simulations and research studies.

NetLogo provides researchers and modelers with an interactive user interface, enabling them to easily prototype and experiment with their models. Its user-friendly interface allows for intuitive visualization of model behavior and dynamics, facilitating the exploration and analysis of simulation results. Moreover, NetLogo offers a powerful feature called BehaviorSpace, which enables the configuration of batch simulations. Researchers can define multiple parameter settings and specify the desired number of replications to be run. BehaviorSpace automates the simulation runs, collecting and logging the results to files for further analysis. This capability allows efficient exploration of different scenarios, sensitivity analysis, and statistical analysis of simulation results. The log files generated by NetLogo simulations can be easily postprocessed using various tools, including popular statistical analysis software such as R or spreadsheet applications such as Excel. Researchers have the flexibility to choose their preferred tool for statistical evaluation, data visualization, and further analysis.

Furthermore, NetLogo supports direct integration with other programming languages, such as Python. This allows researchers to take advantage of the capabilities of Python libraries and tools for data manipulation, analysis, and visualization. The seamless integration between NetLogo and Python [151] or R [381] allows the use of advanced statistical methods, machine learning algorithms, and custom analysis pipelines to gain deeper insights from simulation results. Overall, the combination of NetLogo's log files and its interoperability with other programming languages provides researchers with a comprehensive toolkit for analyzing and interpreting simulation outputs in a flexible and customizable manner.

In NetLogo, simulations are time-based and operate on a discrete scale using ticks. Agents in the simulation are categorized into different types, called "breeds", and these agents can interact with each other in various ways. These interactions can occur through direct proximity to a two-dimensional plane of patches, through connections defined by a network topology, or indirectly through the use of residual information in the environment.

We implemented the SwarmFabSim simulation framework in NetLogo implementing the use case and the modeling approach for swarms in production plants as described in Section 3.2. This framework supports dispatching and scheduling modes, single-lot oriented and batch machines, and an arbitrary number of machine and lot types. SwarmFabSim is a modular system comprising multiple code modules that communicate through

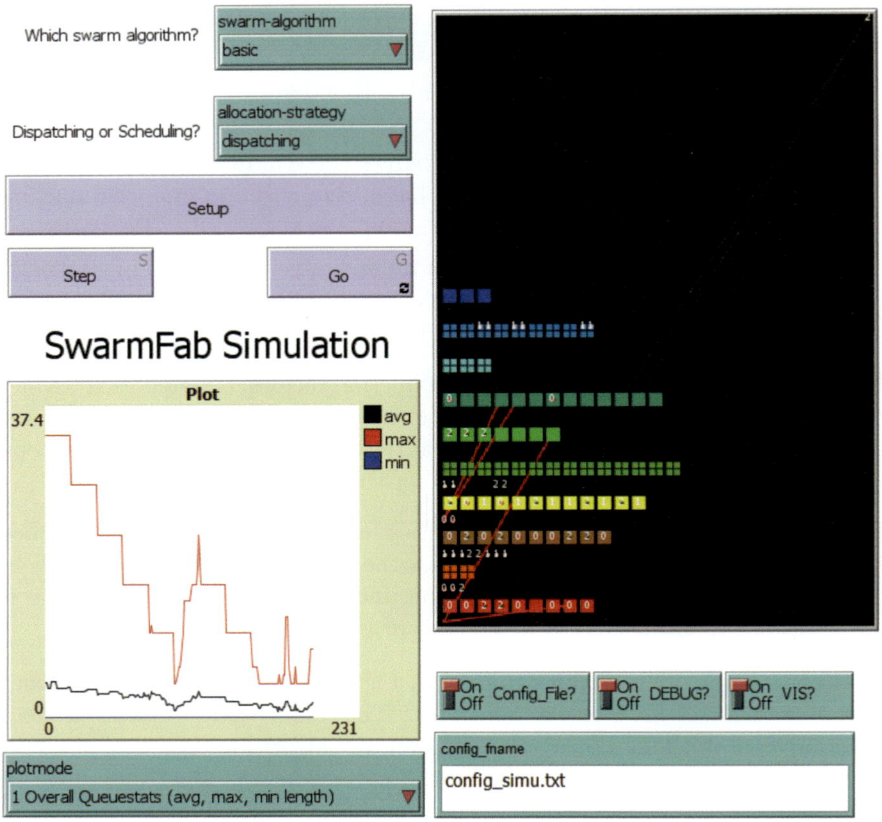

Figure 5.2: SwarmFabSim: Screenshot of the user interface.

a callback architecture. The user interface, as depicted in Figure 5.2, along with the configuration files, enables users to interact and customize the SwarmFabSim framework to align with their specific requirements of the fabrication model.

The simulation scenario is defined using a set of plain text configuration files:

META contains the names of MFILE, RFILE, and LFILE config files. This file bundles the associated config files together. The name of the META config file to be used is set through the "config_fname" input field on the UI or the BehaviorSpace settings.

MFILE is a configuration file that encompasses the machine definitions within the simulation scenario. For each machine type, denoted as m, this file specifies various attributes:

- Process ID: It determines the specific process, represented by P^m, that the machine can perform.

- Number of Machines: It denotes the total count of machines available for this particular type.
- Processing Time: It indicates the time required for the machine to complete a single task.
- Batch Size: This parameter defines the number of lots that can be processed as a batch. If the batch size is set to one, the machine operates in a single-lot orientation.
- Maximum Waiting Time: For a batch machine, this value indicates the longest time the machine will wait for a batch to be filled before starting the process.

RFILE serves as a repository for all production recipes, denoted as R^t, which are utilized for manufacturing various lot types t. These recipes are essentially straightforward lists that outline the sequential process steps, represented by P^m, in the specific order they need to be executed.

LFILE specifies the production quantities for each lot type t based on their corresponding recipe R^t. It defines the number of lots that should be produced for each specific lot type, indicating the desired quantity of output according to the prescribed recipe.

The callback architecture depicted in Figure 5.3 is implemented through the API defined in the file hooks.nls. The main code invokes a hook function contained in hooks.nls whenever an algorithm has the potential to perform an action. Subsequently, the corresponding algorithm function is invoked from the hook function, enabling the desired functionality to be executed. In order to install an algorithm, you need to provide a file named "algorithm-*name*.nls" that implements the algorithm based on the hook API. Additionally, you should add the algorithm to the UI chooser labeled as

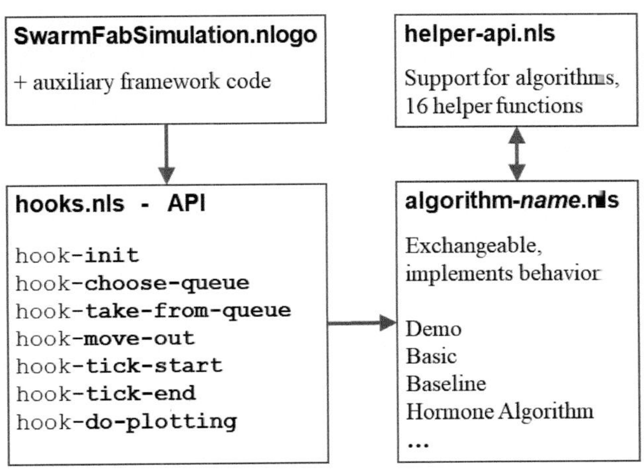

Figure 5.3: SwarmFabSim architecture [390].

"algorithm" to make it accessible for selection. Furthermore, the algorithm functions should be incorporated into the appropriate hook functions within the hooks.nls file for callback purposes. It is also possible for algorithms to utilize helper functions available in the "helper-api.nls" file to streamline their implementation and enhance their functionality. These helper functions offer additional functionality and can be leveraged by algorithms to facilitate their operations.

The reader is considered to be aware that while our approach draws inspiration from semiconductor manufacturing, it can be easily adapted to other industries that employ flexible job-shop scheduling. This adaptability is achieved by modifying the configuration files that define the specific parameters of the given industry setting.

The complete NetLogo source code of SwarmFabSim, along with a collection of configuration files, is openly accessible in a dedicated GitHub repository [389]. Users can freely explore, utilize, and contribute to the development of SwarmFabSim through this open-source platform.

5.3 Physics-based Simulation

Physics-based simulation considers (all) the physical parameters of a robot and the surrounding environment. This type of simulation requires higher processing power; hence, a simulation task becomes more complex when the population of the swarm is increased. Many simulation platforms support physics-based simulation, as mentioned in Table 5.1. ARGoS[1] and Webots[2] are two successful platforms that are widely used for swarm robotics applications. They generate realistic outputs for multi-robot experiments by implementing robot and robot-to-robot physical contact. The main drawback of using the physics-based simulation for large swarm experiments is the computational complexity and limitations resulting from the memory size and processors of the host computer. Although this limits the implementation of large population experiments, there are programming techniques that reduce the issue, e.g., deactivating visualization of experiments. There are many successful implementations of swarm behaviors using physics-based simulation, such as task allocation [48], self-adaptive communication strategy [119], cooperative navigation [95] and swarm controller design [124] in ARGoS; collision-free flocking [29] and formation control [169] in Webots; aggregation [22] and foraging [2] in Stage; and swarm navigation and collision avoidance [264, 309] in Gazebo.

Most simulation software contains a comprehensive list of libraries for commonly used robots; hence, we can easily select a robot from the list and use it. Figure 5.4 illustrates an example of mobile robots available on the Webots platform. Furthermore, most simulation platforms allow us to define a new robot and import the robot's mechanical design, sensory system, and

[1] ARGos website: https://www.argos-sim.info/[Online; accessed: 19-December-2024].
[2] Cyberbotics website: https://cyberbotics.com/[Online; accessed: 19-December-2024].

Figure 5.4: Physics-based models of some of the available robot libraries in Webots software; (left) large and (right) small mobile robots.

physical properties. For example, in [11], a simulated Mona robot model was developed and used to implement an exploration scenario in a swarm system (see Figure 5.5). In this example, infrared sensors, DC motors, LEDs, and power systems were virtualized and modeled in the simulation.

(a) (b)

Figure 5.5: (a) Mona, an open-source, cost-effective robot designed for swarm robotics [17], and (b) a virtual model of Mona in Webots [11].

Another example of importing a new robot into the Webots software is shown in Figure 5.6(b). The Colias micro-robot, widely used for swarm robotic research, is modeled with great detail. The model is equipped with an RGB camera, three IR proximity sensors, and a power system similar to the one developed for the real robot. The detailed model shown in Figure 5.6(c) is mainly for presentation purposes; therefore, the simulator's visualization function must be deactivated if we need to run experiments with a larger number of robots (> 25 robots) on a standard desktop computer.

Physics-based simulations are increasingly used in more complex decentralized swarm systems. Typically they are is used in the form of a feasibility study before the swarm controller is tested with real-world robots. For example, in Hu et al. [169], a framework for a fault-tolerant Search

(a) (b) (c)

Figure 5.6: (a) Colias, an open-source micro-robot developed for swarm robotic applications [19], (b) a virtual model of Colias in Webots and (c) a detailed CAD model of Colias.

& Rescue mission was built with a heterogeneous swarm system. They implemented a complex controller to steer the flock and an autonomous game-theoretic decision-making algorithm. It is obvious that developing such a complex system with a communication network and heavy computation directly on real robots is very challenging. Therefore, the physics-based simulation, in particular Webots, was chosen to test the proposed system by simplifying communication networks and tracking the system using a central observer. Figure 5.7 illustrates the scenario proposed in the Webots simulation. Other similar studies on formation control of multi-agent systems were presented in McCord et al. [241] and in Rekabi et al. [308] that simulated large multi-UAV systems in Gazebo. There are many more examples of multi-robot systems tested in physics-based simulations prior to real-robot implementations. In Aranda et al. [14], a formation control of a multi-UAV was simulated in Cobaye software [396], also a physics-based simulator. Another study in Baumann and Martinoli [30] proposed a modular framework for the navigation of a multi-robot system. The framework was implemented in Webots and the swarm robotics platform Khepera-IV. Moreover, for a multi-robot path planning system in a robotic manufacturing scenario, an optimal sequential task allocation was proposed in Brown et al. [47] which was simulated in Webots as well.

Another research direction in CPS swarms involves multi-vehicle and connected vehicle platoons. Recently, numerous studies have focused on the control challenges of multi-vehicle systems. These strategies are predominantly tested using physics-based simulations. For instance, vehicle fleet target tracking [352] has been simulated in CARLA [93], model predictive control [400] simulated in Prescan/Matlab/V2X, and Unreal Engine [128] has been utilized to simulate overtaking [413] and adaptive cruise control [415] in multi-vehicle systems. Figure 5.9 shows examples of multi-vehicle experiments using physics-based simulations.

Moreover, physics-based simulation are widely used for machine learning in swarm systems. Training a model using deep reinforcement learning is a time-consuming process. The system needs to run for thousands of iterations

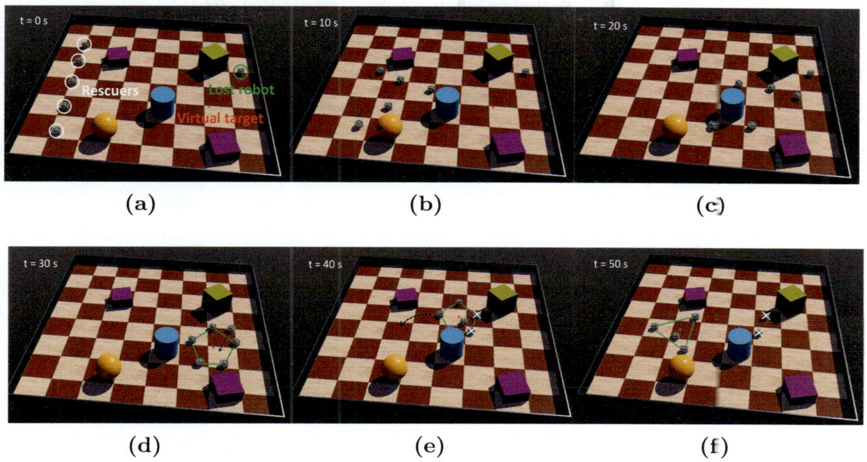

Figure 5.7: The Search & Rescue mission began at $t = 0$ s. Ten seconds later, rescuers were on their way to the virtual target. At $t = 30$ s, they had formed a pentagon around the target. By $t = 40$ s, the lost robot had been contained and was being guided home. Unfortunately, two of the rescuers (marked by White crosses) had stopped working due to an unexpected fault. At $t = 50$ s, the remaining rescuers had created a new triangular formation to guide the lost robot back to the starting point [169].

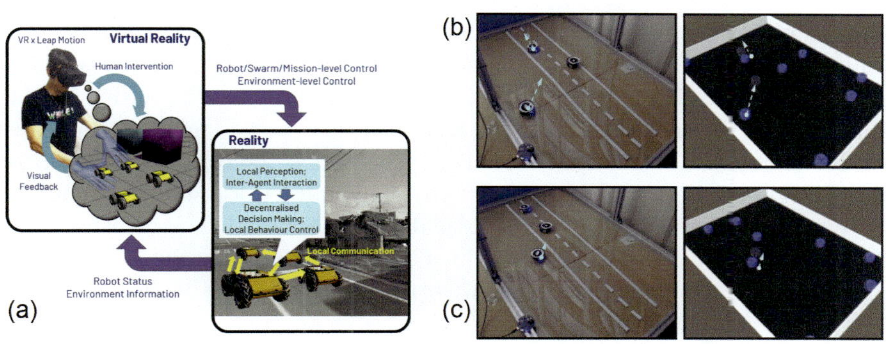

Figure 5.8: (a) The system architecture delineating the proposed human-swarm Interaction utilizing Omnipotent Virtual Giant, as detailed in Jang et al. [178]. Experimental validation showcasing (b) the relocation of holographic objects of robots and (c) the subsequent movement of real robots towards these relocated objects. Dashed arrows indicate the remaining trajectory to the target objects.

to train a good model. Therefore, developing models for swarm systems using real robots can be very difficult and almost impossible due to the autonomy time of the robotic platforms and their limitations in processing and memory units. Researchers tend to use physics-based simulation to train the model before they transfer it to real robots. As an illustration,

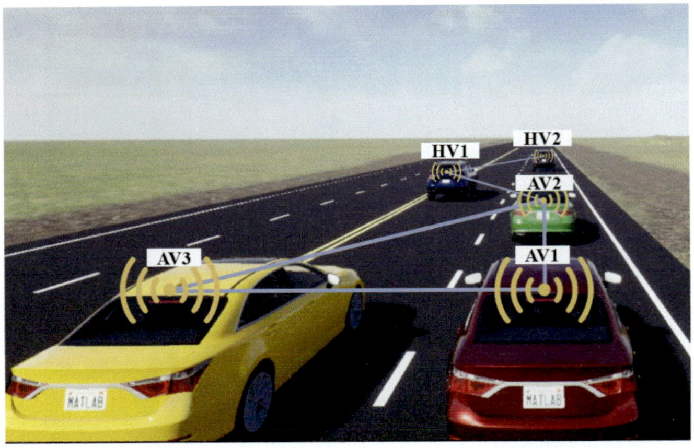

(a)

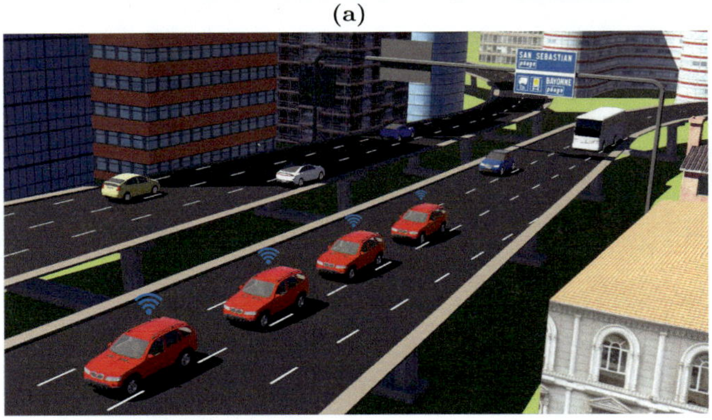

(b)

Figure 5.9: Screenshots of simulation examples of multi-vehicle simulations using (a) Unreal Engine [414]and (b) Webots [168].

Hu et al. [170] established an autonomous exploration task employing a multi-robot system. In this scenario, a model was trained through deep reinforcement learning within the Gazebo simulation environment. Figure 5.10 illustrates the training phase that has been done in Gazebo and real robot experiments with TurtleBot3 robots. A similar study proposed a collective navigation and obstacle avoidance system based on a federated reinforcement learning approach [268]. It was implemented with real robots, TurtleBot2, after the model was developed in Gazebo. In another study, a collision avoidance approach was developed in which robots were trained using deep reinforcement learning in Stage simulation [108]. A decentralized scenario with individual robots that generate paths considering a limited sensory system was investigated. There are many successful examples of using physics-based simulation for robotic swarm systems, such as human-swarm interaction to train artificial neural networks [75] and obstacle avoidance of robot swarms using a trained artificial pheromone system [265, 264].

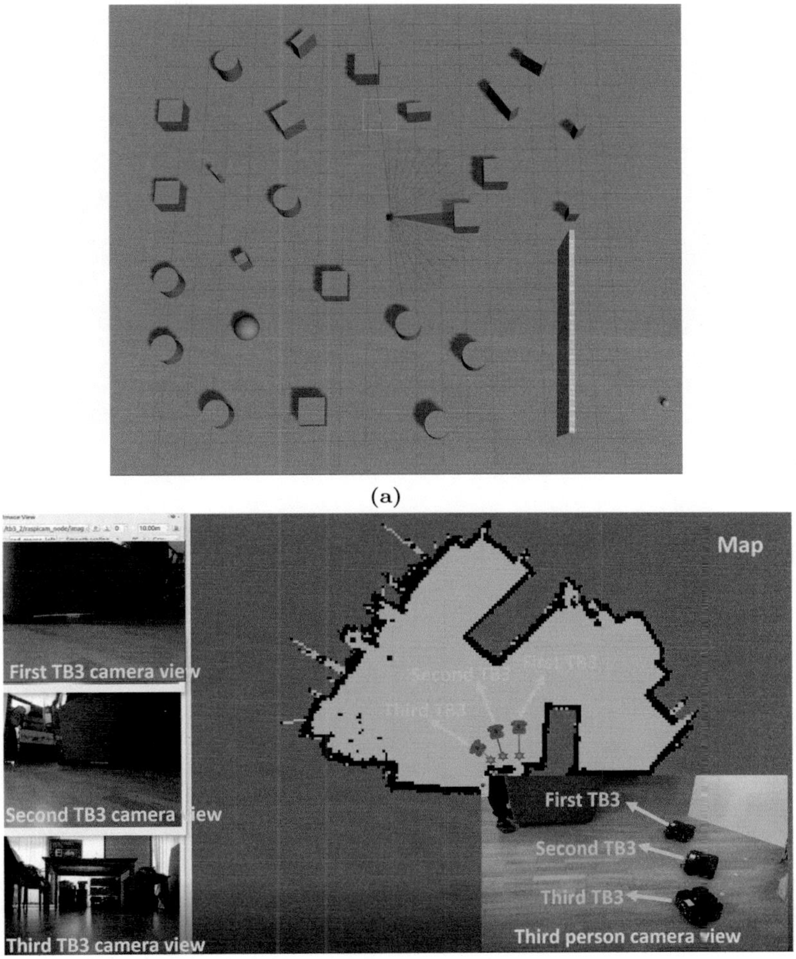

(a)

(b)

Figure 5.10: (a) Gazebo environment for training and (b) real-world experiment using three TurtleBot3 Waffle Pi mobile robots [170].

Physics-based simulations are also combined with real robots in the development of mixed-reality missions. In these settings, a larger population is usually developed in a simulation software, and a few of these robots represent real robots in a real-world setting. A recent work of Jang et al. [178] presents an interesting implementation of a mixed-reality human-swarm interaction where real robots are deployed in the presence of virtual robots in a bespoke developed Unity-based simulation platform. A human operator supervises a swarm by interacting with simulated robots [139]. An additional instance of mixed-reality human-swarm interaction is demonstrated in Patel et al. [284, 283], enabling human interaction with the swarm at both environmental and swarm levels.

5.3.1 BeeGround: A Simulation Platform

BeeGround [229] is an open-source simulation platform developed on the Unity Development Engine [127]. The main purpose of developing BeeGround was to make a fast and reliable swarm simulation accessible for everyone, with the capability of simulating swarm scenarios from a very abstract level to advanced detailed physical models. This allows researchers to increase and tune the level of modeling as needed. With Unity's physics engine and design interface, a plug-and-play package was developed, allowing researchers to implement a swarm scenario and environments of varying sizes, obstacle placements, and swarm populations to develop desired swarm controllers. Different sensors and other modules can be easily added or removed. The actuation mechanism can be altered, or the agent can be swapped out entirely. In addition, Unity supports TensorFlow 2.0 integration that expands the application of BeeGround for machine learning in swarm robotic systems. Figure 5.11 shows a swarm scenario developed using the BeeGround platform implementing a cue-based aggregation with a large swarm of Mona robots.

Before starting a simulation, we can customize the swarm's behavioral and physical parameters. We start by setting up an area where the size, shape, and presence of obstacles can be defined to create a variety of scenarios. For example, when using the cue-based bio-inspired algorithm, BEECLUST, an additional heat map can be loaded that is used by the agents to reference the temperature conditions within a region. The swarm parameters provide users with the capability to define key aspects such as population size, initial agent positions within the arena, and various kinematic constraints. Furthermore, integration packages for ROS-Unity facilitate the incorporation of well-established ROS projects into Unity, allowing for the seamless publication and subscription to topics between the two platforms. Detailed programming in Unity is not expounded upon here, given the extensive library of learning resources within the development engine and the availability of an active community for assistance and comprehension. Instead, we provide an overview of the components necessary to start and operate BeeGround.

Robot Modeling: Unity Engine scripts have two main components that form the basis of the robot's controller: A start function and an update function. The start function is analogous to the initialization phase of the robot and is executed at the start of the simulation. The update function is similar to a `while` loop and is continually executed throughout the simulation. The robot's behavior is determined within this update function.

We can define a robot by either importing a standard Unified Robotic Description Format (URDF) file or constructing one within the Unity Environment. Rigid body components can be added to the robot, which will be affected by the physics engine. This allows for the application of forces, torques, and collisions within the engine's fixed update function. Joints can also be implemented, which enables the creation of wheels and robotic limbs. Moreover, sensors such as cameras and range finders can be created using the tools provided by the Unity development engine.

Arena Configuration: The initial step in configuring the arena is to specify its size in standard units. BeeGround will then create a walled-off area with the given dimensions. Afterwards, the user can input an occupancy grid to place a cubic unit of obstacles in the environment. Furthermore, Unity provides other assets that can be used to create more intricate obstacles if desired.
Swarm Parameters: Bees, which are models of robots, can be created from the beginning with adjustable parameters, ranging from abstract to physics-based simulation. Instead of the Bee agent, custom robot models can be used for swarm generation. Furthermore, BeeGround allows the swarm population and placement to be specified, thus enabling some extraordinary testing scenarios.
Simulation Parameters: For ease of use, the length of a simulation and the number of repetitions can be specified in this section. Additionally, the speed of the simulation can be adjusted while it is running. This feature offers great versatility for running extended experiments with multiple repetitions that are necessary for statistically analyzing the outcomes.
Bio-inspired-specific Parameters: The environmental conditions, such as humidity, temperature, light, and more are essential for bio-inspired swarm robotics scenarios. To create a realistic simulation platform for robotics swarm applications, these environmental properties must be incorporated into the simulator. For example, a heat map was included in Wang et al. [401] to test the honeybee aggregation algorithm in a complex environment. This heat map was made up of an array of temperatures that the agents used as a reference for their waiting times. We can use multiple layers of these arrays to introduce environmental properties from different sources and models.

We can also create dynamic models of the environmental conditions that vary over the course of an experiment. These environmental characteristics can interact with the robots, for example, to create a bio-inspired pheromone communication system. To log swarm experiments, we record various parameters of the agents, such as position and orientation, every second to observe the behavior of the swarm over time. The output logs can be tailored to the user's requirements, as other parameters, such as velocities and rotations, are accessible through the Unity interface.

As previously mentioned, faster simulations can be achieved by reducing or deactivating the physics, although this comes at the cost of accuracy. This allows us to simulate swarm behaviors with thousands of robots, as illustrated in Figure 5.12. This capability of running experiments with very large swarm populations opens up new research directions that have not been explored before. For example, in Kiszli and Arvin [207], a bio-inspired cue-based aggregation was studied with a large swarm robotic system. It was found that the original goal of BEECLUST aggregation could not be generalized as a result of a barricade effect observed from experiments with large swarms.

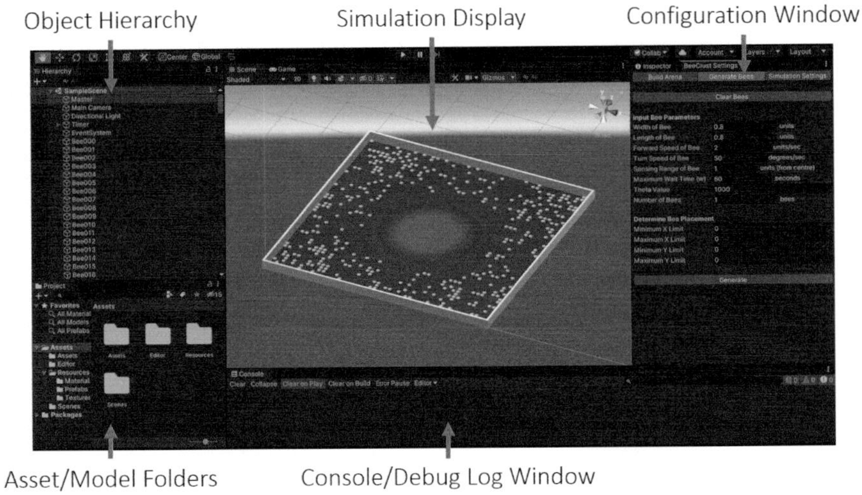

Figure 5.11: BeeGround UI featuring arena with a gradient cue with a large size swarm. The *Object Hierarchy* provides an overview of all the elements in the environment, such as robots, walls, obstacles, and cues. The *Simulation Display* is a graphical representation of the simulation. The *Configuration Window* allows users to adjust the settings for the BeeGround simulation, including the agents and their behavioral parameters. The *Asset/Model Folders* is a library of all the assets related to the simulation. Finally, the *Console/Debug Log Window* tracks all errors or debugging information that occur during the simulation [229].

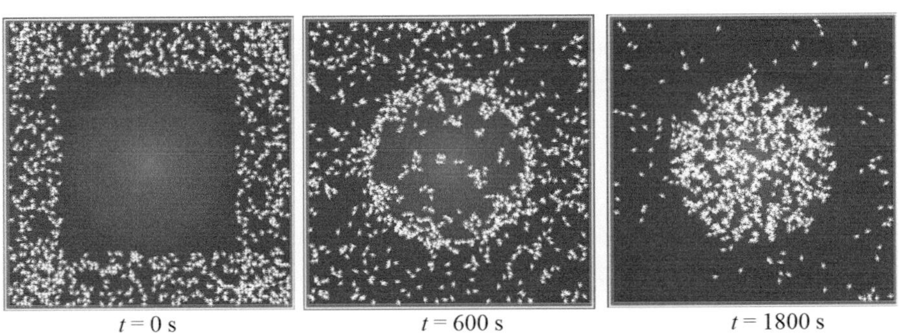

Figure 5.12: An example of BEECLUST aggregation in the BeeGround with 1000 robots in an environment with a single gradient cue (from [229]).

Chapter 6

Swarm Robotic Platforms

Numerous robots have been developed for use as swarm robotic platforms, such as Spiderino [181] and Colias [19], each tailored for a particular task mimicking bio-inspired behaviors and scenarios. Additionally, various general-purpose mobile robots like e-puck [249] and Thymio [250], have been widely employed in robotic swarm applications. Swarm robotic platforms typically share common features such as small size, low-cost and a preference for open-source hardware and software. Furthermore, it is crucial to have versatile robots that are easy to program and modular, enabling the re-utilization of robots in multi-purpose tasks. We will examine the fundamental capabilities and explore the basic functionalities of a swarm robotic platform chosen for swarm applications.

6.1 Sensors

To interact with the environment and other robots, a reliable sensory system is essential for a swarm robotic system. Standard sensors such as short- and long-range IR and ultrasonic proximity sensors, as well as low-resolution cameras, are commonly used. Long-range IR sensors are especially popular because of their simplicity, as they operate by driving a photodiode to transmit and reading IR values from a phototransistor as a receiver. These sensors enable robots to identify other members of the swarm and distinguish themselves and others, from obstacles or other elements in the environment. This capability is important for the localization of the swarm [233]. The range of these sensors can vary from 10 cm to a few meters, depending on the IR radiance intercity (measured in mW/sr, milliwatts per steradian). Reflected IR from obstacles, such as walls or other objects follows the fundamental principles of

electromagnetic radiation and can be mathematically modeled as,

$$s(x,\theta) = \frac{\alpha \cos\theta}{x^2} + \beta, \tag{6.1}$$

where $s(x,\theta)$ represents the recorded value from the sensor, with x denoting the distance of the obstacle from the sensor, and θ indicating the angle of incidence with the surface. The variables α and β encompass various parameters, including the reflectivity coefficient, the output power of emitted IR, the sensitivity of the sensor, and the ambient light effect. These parameters are typically estimated empirically. Consequently, white and dark surfaces exhibit different ratios of reflection and absorption of IR radiations. This distinction is a critical consideration when devising an experimental setup for a swarm system. In swarm robotics, understanding the surrounding environment is a crucial requirement for a robot. They need to detect neighboring robots before an interaction starts; hence, several sensors are placed around a robot. As an example, Figure 6.1 illustrates the sensory reading of the Colias micro-robot that detected another robot in front. The varying intensity of the IR readings from multiple receivers enables the robot to estimate the relative position and orientation of neighboring robots using the following equation:

$$\phi = \text{atan}\left(\frac{\sum_{i=1}^{n_s} \hat{s}_i \sin(\gamma_i)}{\sum_{i=1}^{n_s} \hat{s}_i \cos(\gamma_i)}\right), \tag{6.2}$$

where ϕ is the estimated bearing of a neighbor, n_s is the number of sensors used by the robot, γ_i is the angular distance of the ith sensor with respect to the robot's head, and $\hat{s_i}$ is the IR intensity received by the sensor i. By employing simple data analysis on the robot's microcontroller, the robot can detect neighboring robots and estimate their position and bearing.

In addition to IR proximity sensors, existing swarm robotic platforms are also equipped with various off-the-shelf sensors, They have been used as cue markers which measure, which measure ambient light illumination have been particularly popular. They have been used as cue markers in in honeybee aggregation [329] and to simulate pheromone trails in ant-like exploration [18]. The more sophisticated version can also read RGB colors (as shown in Figure 6.2), which facilitates implementing multi-layer pheromone systems, as demonstrated by Liu et al. [231]. More details about the pheromone system and its implementation for swarm robotics will be discussed later in this chapter.

6.2 Actuators

Swarm robotic platforms employ various types of actuation mechanisms that have been used to actuate swarm robotic platforms. For instance,

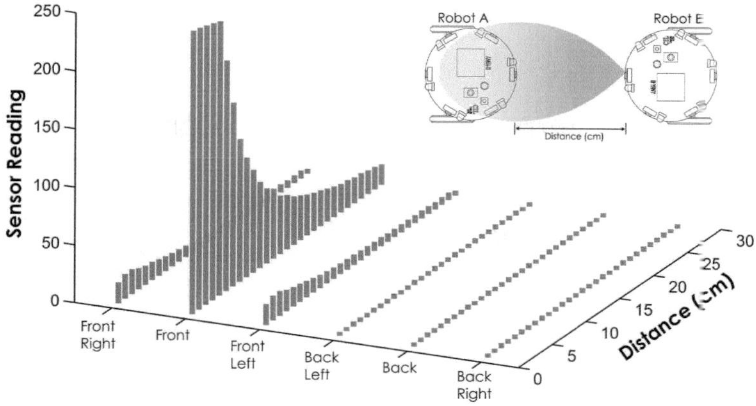

Figure 6.1: An example of a swarm robotic platform, equipped with six IR sensors, where Robot A receives IR from Robot B that has a 0° orientation (from [19]).

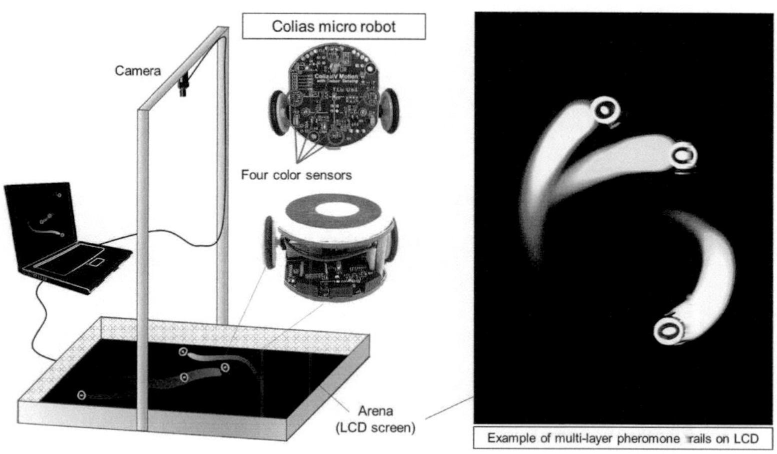

Figure 6.2: An example of a multi-layer pheromone system (from [231]).

Spiderino [181] uses six legs for walking on the floor, Kilobots [316] move on smooth surfaces by vibrating their three legs, UAVs in the CoCoRo project [330] used thrusters for underwater swimming, and drones in Crazyswarm [293] use propellers for flight. However, due to their simplicity of control, wheeled robots are the most common actuation mechanism in swarm robotics. There are also hybrid mechanisms such as tracks and wheels developed by Mondada et al. [252], so called treels, that make navigation simpler and smoother.

In terms of wheeled robots, there are various types of wheels configurations such as differential drive (two and four wheels), car-type steering, omnidirectional and synchronous drive, among others. Due to the mechanical and control simplicity of the two-wheel robots, many swarm

robotic platforms have two wheels with a differential drive steering mechanism, which controls the trajectory of the robot by controlling the left and right wheel independently.

6.3 Communication

There are various types of communication used for swarm robotic platforms, direct communications using, e.g., IR, Wi-Fi, Bluetooth as communication technology to send specific messages, and implicit communication, communication over the environment, for example, pheromone-based. The direct form of communications is implemented by: (i) One-to-one, where a robot can directly talk to another robot and share information, (ii) one-to-many, where a robot can broadcast its messages to many individuals (leader to followers), (iii) many-to-one, where several robots communicate with an individual (followers to leader), and (iv) many-to-many, where swarm members communicate with each other without having a priority between them. As passing messages are always very specific to the underlying communication technology, we will focus on stigmergic, pheromone-based communication in the next section.

6.3.1 Pheromone-based Communication

Pheromones are a chemical substance that are detected by members of the same species, causing them to act in a certain way [194]. These substances play a vital role in facilitating communication among a diverse array of organisms, ranging from yeast and insects to mammals. Vertebrates have been discovered to use pheromone-based communication. Research has revealed that humans can experience physiological and psychological reactions due to chemosignals [263]. Many animals rely on pheromones as their main form of communication, but, to the best of our knowledge and the swarm perspective, social insects make the most out of this method. Pheromones enable extensive communication among a group of insects, creating a shared memory-like effect. Moreover, the utilization of pheromones serves to optimize the collective behavior of social insect groups. For instance, ants employ pheromones and feedback systems to efficiently locate the shortest route from their nest to a food source.

As an example of pheromone-based communication in social insects, Monomorium pharaonis called Pharaoh's ants, which are usually found in human habitats, use multiple types of pheromones that are important for their foraging behavior in dynamic and competitive environments [177]. For example, they generate a complex combination of pheromone trails between the food source and the nest using three distinct types of pheromones (i) non-volatile attractive pheromones, (ii) volatile attractive pheromones, and (iii) repellent pheromones. In another example, bumblebees, Bombus hortorum, leave chemical cues on flowers that allow detection and avoidance of recently depleted flowers [103]. Similarly, chemical cues enhance the efficiency of bee

colonies' foraging behaviors, as they prevent meaningless visits to depleted flowers. Another type of pheromones in nature are queen pheromones that characterize the queen and other members, which are essential to maintain the colony. For example, if the queen fails for any reason, including viruses and pesticides, the secretion of the queen pheromone in the colony decreases, causing the colony to collapse [365].

Pheromones in Robotics

Pheromone-based communication is one of the bio-inspired communication mechanisms widely used in swarm robotics. It requires a simple capability for an individual robot since it only needs local sensing. Additionally, the environment plays the role of memory; hence, individuals only need to have limited local memory. Moreover, we can optimize the performance of a swarm system using a combination of multiple types of pheromones and feedback mechanisms. And, most importantly, the system is fully decentralized without requiring a central control.

One of the first works inspired by pheromone communication was introduced by Russell [318]. This work used cinnamon camphora, known as Camphor, as the implementation of trail pheromones for trail following behaviors for robotic systems, which embodied odor release and detection functionalities. It showed the applicability of pheromone-based communication in robotic systems. In another research work [125], they used ethanol to simulate a pheromone for the robotic system.

Furthermore, numerous studies have suggested robotic systems that leverage Radio-Frequency Identification (RFID) tags as a medium for pheromonal communication [7, 159, 202]. RFID tags were attached to the floor where the robots operate. The tags store data transmitted by the robots passing above and eliciting the corresponding behavior depending on its nature. This principle works like pheromones in nature. Recent studies have involved virtual environments in implementing artificial pheromones for the communication of swarm robotic systems. As an illustration, Campo et al. [54] introduced a mechanism for path selection in a foraging robot swarm employing virtual ants. In this approach, robots engaged in local transmission and reception of messages. Although this work implemented virtual pheromones only within robots, the other works created a virtual map to mark the deposited pheromones accessible to all robots [307, 317]. The robots are outfitted with virtual sensors, accessing shared virtual environments among all robots, which incorporate overhead tracking and control. There are many interesting implementations of artificial pheromone communications, for instance, Kilogrid [391] developed for Kilobot micro-robot, Phormica [321] implemented with e-puck. Several researchers have implemented pheromone-based communications by various means. Garnier et al. [131] simulated pheromones by projecting light from a ceiling mounted video projector.

Figure 6.3: On the left is the Colias micro-robot and on the right is the Colias bottom board, featuring pheromone sensing capability. Various modules of Colias include A) main processor, B) IR proximity sensors, C) digital camera, D) micro-motors with gearhead, E) 22 mm wheels, F) pheromone detectors (light intensity sensors), G) battery recharging unit, H) main switch, and J) ISP programming port (from [266]).

COSΦ is an open-source artificial pheromone system for swarm robotics that uses light-based pheromone trails [18]. The utilization of an LCD screen served as the arena for robots to engage with pheromones, visually represented as illuminated spots on the screen. Through a tracking system facilitated by a camera positioned above the arena, the system continually updated the robots' position, orientation, and ID, generating virtual pheromones in response. COSΦ offers several advantages: (i) It boasts a notably high resolution for pheromone implementation in comparison to other approaches, (ii) its highly flexible nature allows for the implementation of diverse environments, including variations in pheromone trail thickness, evaporation rate, and diffusion, and (iii) it employs a cost-effective configuration featuring a basic digital camera [212]. A flat LCD screen enhances the accessibility for a wide range of researchers. The Colias microrobot [19], shown in Figure 6.3, was deployed for the implementation of artificial pheromone communication. It has been extended in several studies [267, 231, 266] to include diffusion, wind effect, multi-layer pheromones, which are phenomena almost ever present in nature. Figure 6.4 illustrates a cue-based aggregation scenario in the presence of pheromones implemented by an artificial pheromone communication.

Artificial Pheromone System

The artificial pheromone system COSΦ comprises two integral components: A pheromone system and a tracking system. The pheromone system calculates and displays the pheromone on a flat LCD screen that is placed horizontally, representing the environment in which the robots are operating [18]. The tracking system tracks the robots and sends their data to the pheromone system [212]. An example of an artificial pheromone environment with real-world mobile robots is shown in Figure 6.5.

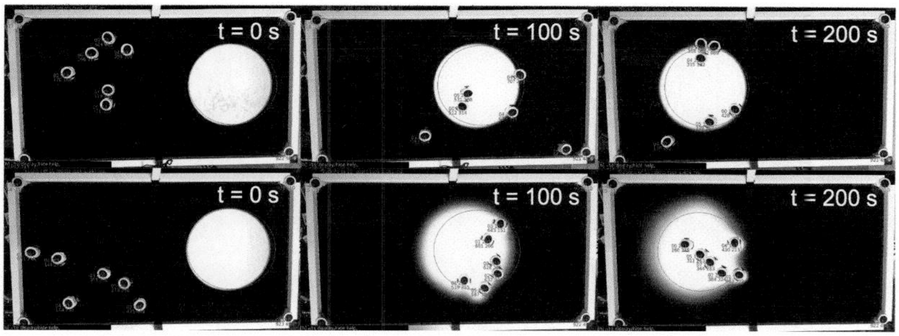

Figure 6.4: The first row presents sample experiments conducted without diffusion and with fast cue speed, devoid of pheromone injection. In contrast, the second row illustrates sample experiments conducted with diffusion and fast cue speed, incorporating pheromone injection, captured at time instances $t = 0$ s, $t = 100$ s, and $t = 200$ s from left to right (from [266]).

The pheromone system models multiple types of artificial pheromones and their interactions simultaneously. The brightness of a grayscale image is represented as $\mathbf{I}$[1], which is a two-dimensional matrix with the size of the resolution of the LCD screen. It is determined by $\mathbf{\Phi}$, which is a two-dimensional matrix of the same size as $\mathbf{I}$ that represents the intensity of the pheromone. Each element of $\mathbf{I}$ is equivalent to the brightness of the corresponding pixel. The brightness of the image at position (x, y), $\mathbf{I}(x, y)$, is expressed with

$$\mathbf{I}(x,y) = \sum_{i=1}^{n} c_i \mathbf{\Phi_i}(x,y). \tag{6.3}$$

The intensity of the ith pheromone at position (x, y) is represented by $\mathbf{\Phi_i}(x, y)$, and c_i is the influence of the ith pheromone on the screen. $\mathbf{I}(x, y)$ is calculated by multiplying $\mathbf{\Phi_i}(x, y)$ and c_i and summing them up, which means that n pheromones can be overlapped. To illustrate how the model works, the combination of three different pheromones with different effects on the screen can be displayed on a single-pixel after the calculation of $\mathbf{I}(x, y)$ using Equation 6.3.

During the operation of the system, the amount of pheromones released on the screen is continuously adjusted at discrete intervals. The intensity of the modified pheromone is expressed with

$$\begin{aligned} \mathbf{\Phi}_i^{k+1}(x,y) = &- \mathbf{u} \cdot {}^{grad}\mathbf{\Phi}_i^{k}(x,y) - \frac{\ln(2)}{e_{i\Phi}} \mathbf{\Phi}_i^{k}(x,y) \\ &+ \kappa_i \nabla^{grad}\mathbf{\Phi}_i^{k}(x,y) + \iota_i(x,y). \end{aligned} \tag{6.4}$$

In the given expression, $\Phi_i^{k+1}(x, y)$ denotes the intensity of the ith pheromone at discrete time $k+1$, while $\Phi_i^k(x, y)$ represents the intensity

[1] Bold font parameters in this section indicate two-dimensional vector.

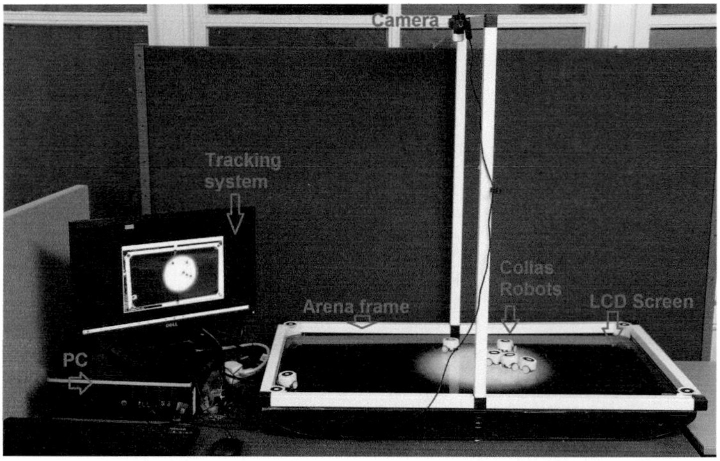

Figure 6.5: The experimental setup employed for the pheromone system in Na et al. [266] encompasses a PC dedicated to tracking robots and generating pheromones, a digital camera for monitoring robot positions, a horizontally positioned 42" LCD screen, an aluminum frame encircling the arena, and the presence of Colias mobile robots.

of the ith pheromone at discrete time k. The quantity ${}^{grad}\mathbf{\Phi}_i^k(x, y)$ is a two-dimensional vector describing the gradient of pheromone intensity at the position (x, y) and is defined mathematically by Equation 6.5. The symbol $\mathbf{u}$ signifies the velocity vector responsible for linearly shifting the pheromone across the arena. The term $e_{i\Phi}$ governs the evaporation rate of the ith pheromone and is characterized by its half-life, κ_i denotes the diffusion constant of the ith pheromone, and $\iota_i(x, y)$ corresponds to a newly injected pheromone at the position (x, y) on the screen. The spatio-temporal development pheromone intensity model is derived from a simplified version of the Navier-Stokes equation, representing the fluid flow model.

$$
\begin{aligned}
{}^{grad}\mathbf{\Phi}_i^k(x, y) = & \frac{\mathbf{\Phi}_i^k(x+1, y) - \mathbf{\Phi}_i^k(x-1, y)}{2}\mathbf{i}+ \\
& \frac{\mathbf{\Phi}_i^k(x, y+1) - \mathbf{\Phi}_i^k(x, y-1)}{2}\mathbf{j}.
\end{aligned}
\tag{6.5}
$$

By recalculating Equation 6.4 for each value of (x, y), it is possible to compute new pheromone intensities from their prior states, which consequently determine the new pixel values of the grayscale image displayed on the screen.

The parameters on the right-hand side of Equation 6.4 can be divided into two categories: Environmental effects and pheromone injection. Environmental effects, such as evaporation rate e_Φ, diffusion constant κ, and velocity vector $\mathbf{u}$ (where i is omitted to generalize), have a constant influence on the pheromone released in the arena while the system is running. In contrast, the injection of pheromone $\iota(x, y)$ only affects the intensity of

pheromone in the arena when certain conditions are met, such as when robots stop. Under these conditions, the pheromone is injected in a circular shape with a given intensity. The injection of pheromone $\iota_i(x, y)$ is defined as

$$\iota_i(x, y) = \begin{cases} s_\Phi, & \text{if } \sqrt{(x - x_r)^2 + (y - y_r)^2} \leq l_\Phi/2 \\ 0, & \text{otherwise} \end{cases} \tag{6.6}$$

where (x_r, y_r) respectively represent coordinates of a robot and (x, y) the position of the pheromones in the arena, s_Φ is the intensity of injected pheromone at a time, and l_Φ is the diameter of injected pheromone. In a circle with a diameter of l_Φ, where the center of the circle is positioned at the robot's coordinates (x_r, y_r), pheromone is uniformly injected at a rate of s_Φ.

Environmental Effects on Pheromones

The versatility of the system is enhanced by environmental effects, which enable it to replicate realistic conditions that influence the spread of pheromones in the environment over time.

Evaporation is the transformation of a liquid's surface into a gas. Volatile chemicals, such as pheromones, evaporate as well. In various studies on the kinetic properties of pheromones, the half-life of the pheromone is used as a measure [398]. This is the amount of time it takes for the pheromone to reduce by half, denoted as $e_{i\Phi}$ in Equation 6.4.

Diffusion is the process by which molecules move from a region of higher concentration to a region of lower concentration. Incorporating diffusion into the pheromone system is essential for realism, as the diffusion of pheromones significantly influences swarm behavior. Therefore, instead of using the mathematical definition of diffusion, $\kappa_i \bigtriangledown^2 \Phi_i(x, y)$, Gaussian blur is employed to simulate diffusion. This approach has the benefit of being able to replicate faster diffusion with fewer computational resources. The intensity of pheromone at the position (x, y) after application of the Gaussian blur is given by

$$\mathbf{\Phi_i^{k+1}}(x, y) = (\omega * \mathbf{\Phi_i^k})(x, y) = \sum_{s=-a}^{a} \sum_{t=-b}^{b} \omega(s, t)\mathbf{\Phi_i^k}(x - s, y - t), \tag{6.7}$$

where $\Phi_i^{k+1}(x, y)$ is the intensity of the ith pheromone at the discrete time $k+1$, $\Phi_i^k(x, y)$ is the intensity of the ith pheromone at the discrete time k and ω is a two dimensional kernel matrix with a size of $(2a + 1) \times (2b + 1)$ defined as

$$\omega(x, y) = \frac{1}{2\pi\sigma^2} e^{-\frac{x^2+y^2}{2\sigma^2}}, \quad w \in \mathbb{R}^{2a+1\times 2b+1}, \tag{6.8}$$

where σ is a standard deviation of the Gaussian distribution of the kernel matrix. Equation 6.8 indicates that the elements of ω are determined by

the Gaussian distribution. This blur technique is advantageous due to its computational efficiency, as well as its capacity to provide more intuitive control of the diffusion rate and area. Integrating Gaussian blur into the pheromone system aligns with diffusion properties for two key reasons in the context of our study: (i) Gaussian blur results in a decrease in the higher intensity of pheromone and an increase in the lower intensity, and (ii) the total pheromone quantity remains constant following each computational step.

Pheromone Shift changes the position of the released pheromone in real and dynamic settings. Advection, which is the movement of any fluid, such as air, that carries pheromone from one place to another, is a natural cause of this shift. The flow of the released pheromone in the same direction is represented by $\mathbf{u} \cdot \nabla \mathbf{\Phi}(\mathbf{x}, \mathbf{y})$. The two-dimensional velocity vector is expressed with

$$\mathbf{u} \cdot \nabla \mathbf{\Phi}_{\mathbf{i}}(x, y) = u_x \cdot \frac{\partial \mathbf{\Phi}_{\mathbf{i}}(x, y)}{\partial x} + u_y \cdot \frac{\partial \mathbf{\Phi}_{\mathbf{i}}(x, y)}{\partial y}, \tag{6.9}$$

where u_x, u_y respectively represents the speed along x-axis and y-axis.

The use of pheromones in swarm systems as a communication mechanism offers scalability of collective behaviors without the need for direct communication between agents. For example, a single ant and a large swarm of ants can use the same pheromone trail to forage without increasing communication costs. This technique has found application in diverse research domains, including optimization [87], vehicle routing [361], and robotics [54, 391, 321, 266]. However, there are still some challenges to overcome before pheromone-based communication can be widely adopted, such as designing controllers for individual agents that can maximize the performance of the entire swarm [91].

Swarm robotic systems that employ pheromone-based communication demonstrate advanced collective behaviors suitable for real-world applications. For example, in Alfeo et al. [7], a vehicle swarm effectively manages waste in a simulated urban setting, surpassing traditional methods. Furthermore, pheromone-based communication is proposed as an optimal solution to manage large-scale autonomous vehicle traffic with a focus on avoiding collisions [265]. Such systems have also been adapted for urban construction scenarios [9]. Beyond these studies, numerous other practical applications of pheromone-based communication are emerging, where complex coordination of large-scale swarm robotic systems are required.

6.4 Swarm Robotic Research Platforms

The collective behavior of a swarm emerges from interactions between simple robots that simultaneously has an indirect relationship with the behavior of each individual robot of the swarm. It is observed that a simple modification of an individual's behavior (such as velocity and sensitivity) could result in a significant positive or negative impact on the collective behavior of the swarm

system. These can happen due to both hardware and software heterogeneity. The robotic platforms developed for a swarm system must be able to reliably imitate swarm behaviors which are mainly found in nature, such as birds flocking [385] and honeybees aggregation [329]. The robots must be designed with a compact size allowing implementation of large-scale swarm behaviors in the laboratory settings at low cost. To implement a variety of swarm behaviors on the individual robots, the design of the the robotic platform mechatronics must be versatile up to a specific level to (i) simplify replication and (ii) ensure platform homogeneity.

In the subsequent sections, we present a compilation of swarm robotic platforms utilized in research and education (refer to Table 6.1). These platforms have been specifically designed for swarm robotics applications or are commonly employed to implement swarm behaviors. The table is structured as follows: Entries are categorized based on the environment in which they operate, terrestrial, aerial, aquatic, and/or space. Each entry includes information on the type of application, the project or product name, the robot type, the swarm size (typically reflecting the number of robots used in the referenced evaluation), and the fundamental swarm behaviors corresponding to the definitions outlined in Section 4.1. We differentiate between various types of swarm robotic platforms including UGV, UAV, Unmanned Surface Vehicle (USV), or Unmanned Underwater Vehicle (UUV). The table and the explanation is an updated and extended version from Schranz et al. [337]. Be aware that other, probably more sophisticated robotic research platforms exist, which are not included. In this table and corresponding sections we focus only on platforms developed with the intention of using them in swarm robotic applications, specifically for research and education.

6.4.1 Terrestrial

The probably best known swarm of robots is the Kilobots swarm [316]. Kilobots have a diameter of only 33 mm, they move with the help of vibration motors, and the communication is implemented with IR light mounted on the bottom of the robot to reflect off the ground. Their most famous application is self-assembling, where they form different shapes using 1,024 Kilobots [412]. The Kilobot is available open-source[2] or commercially at K-Team[3].

Another widely used swarm robotic platform is Jasmine. Also this plattform is available open-source[4]. It was mainly built for large-scale swarm robotic experiments, where each robot is equipped with a set of sensors for touch, proximity, distance, and color. A similar aim regarding large-scale swarms was also actively pursued with the swarm robotic platform Alice [55]. A bunch of additional sensors, such as linear cameras extend the

[2]Kilobot website: `http://www.kilobotics.com/` [Online; accessed 26-March-2023].
[3]K-Team website: `https://www.k-team.com/` [Online; accessed 26-March-2023].
[4]Jasmine website: `http://www.swarmrobot.org/` [Online; accessed 26-March-2023].

basic capabilities. A series of research platforms building upon each other is given with AMiR [20], Colias [19] (open-source[5] and commercially[6] available), and Mona [24] (open-source[7] and commercially[8] available). The platform R-One [242] is also designed for usage as a swarm robotic platform. As a support for close to swarm intelligence experiments it uses a camera tracking system for ground-truth localization and a server-side located software connecting all the pieces. The Elisa-3 swarm robotic platform, open-source and commercially[9] available, uses an Arduino microcontroller, including a large variety of sensors including eight IR proximity sensors, three-axis accelerometer, and four ground sensors. The swarm robotic platform has the capability to recharge autonomously using a charging station. The communication in the swarm is performed either via IR or radio. The Khepera IV [355] is designed for any indoor lab application. The technical features including a Linux core, color camera, WLAN, Bluetooth, USB Host, accelerometer, gyroscope, microphone, loudspeaker, three top RGB LEDs, and improved odometry makes this swarm robotic platform a compact and complete research platform for swarms in different scenarios. The Khepera IV is commercially available at K-Team[10]. The GRITSbot [288] is the open-source[11] swarm robotic platform used in the Robotarium[12] at Georgia Tech, Atlanta. The Robotarium is a physical environment that provides remote access to a large swarm of robots. Scholars, researchers and anybody interested in this topic can upload code to run experiments remotely. Coming with features like automatic registration of robots with a server, autonomous charging, wireless code upload to the robots, and automatic sensor calibration makes the Robotarium attractive for remote research experiments. The swarm robotic platforms in the Robotarium use wheeled locomotion and are equipped with a set of different sensors, for example, distance and light sensors.

The e-puck robot [249], together with its successor e-puck2, represent an educational and research robot designed to ease programming and control of a robot's behaviors. It uses diverse sensors, such as IR proximity sensors, a CMOS camera, and a microphone. The e-puck is available open-source[13] or

---

[5]Colias open source website: `https://github.com/Farshad-Arvin/Mona-Platform/blob/master/Mona-Test.ino` [Online; accessed 26-March-2023].

[6]Colias commercial website: `http://www.visomorphic.com/` [Online; accessed 26-March-2023].

[7]Mona open source website: `https://github.com/MonaRobot` [Online; accessed 26-March-2023].

[8]Mona commercial website: `https://ice9robotics.co.uk/` [Online; accessed 26-March-2023].

[9]Elisa-3 website: `https://www.gctronic.com/doc/index.php/Elisa-3` [Online; accessed 26-March-2023].

[10]K-Team website: `https://www.k-team.com/` [Online; accessed 26-March-2023].

[11]GRITSbot website: `https://github.com/robotarium/gritsbot_2` [Online; accessed 26-March-2023].

[12]Robotarium website: `https://www.robotarium.gatech.edu/` [Online; accessed 26-March-2023].

[13]e-puck open source website: `http://www.e-puck.org/` [Online; accessed 26-March-2023].

commercially at GCtronic[14]. An extension of the e-puck presents the Xpuck. It has additional aggregated raw processing power (as used in modern mobile system-on-chip devices) of two teraflops on board. Thus, higher-individual robot computation can be achieved, for instance, image processing using the ArUco Marker tracking [187]. The Thymio II robot [312] is built on understanding of programming and robotic concepts by using a wide range of sensors, e.g., temperature, IR distance, accelerometer, and microphone. It supports both, visual and text-based programming using Blocky. The Thymio II is available both open-source and commercially[15]. Another platform for open-source[16] swarm robotics development for education and research purposes is called Pheeno [406]. Custom modules allow the adaption of the platform according to the user's needs. Using IR sensors, it interacts with the environment. The Spiderino platform [181] is a six-legged open-source[17] robot with spider-like locomotion. The basis forms a hexpod toy that is additionally enhanced with a PCB on top. This Printed Circuit Board (PCB) includes an Arduino microcontroller, a WLAN module, and several reflective IR sensors to allow robot control.

In the project I-Swarm (Intelligent Small-World Autonomous Robots for Micro-manipulation) the goal is to develop micro robots to form a swarm. The robot has a small size of only $3 \times 3 \times 3\,\text{mm}^3$, it is battery-free powered with solar energy, uses vibration motors to perform locomotion, and IR transceivers for communication [347]. The goal was to build a swarm of 1000 robots [347]. The robot's prototypes are exhibited in the technical museum in Munich, Germany.

The idea behind the open-source swarm robotics platform Zooids[18] is different: It handles both the interaction among the robots and the display/interaction to human operators. Thus, it offers a new class of human-computer interfaces. The control of the swarm is performed through light patterns projected from the top using an overhead projector [223]. The APIS[19] (Adaptable Platform for Interactive Swarm) has multiple components: The swarm robotic platforms, the test environment for the swarm including the necessary infrastructure, and the simulation framework [82]. The focus is on experiments related to human-swarm interaction. For such interactions, the robotic platforms are equipped with an additional OLED display and a buzzer. The Wanda [201] platform follows a special composition that could be used, for instance, to clean up the environment with a swarm. In addition, the authors implemented a robot-specific tool chain from design, simulation to

[14]e-puck commercial website: `https://www.gctronic.com/e-puck.php` [Online; accessed 26-March-2023].

[15]Thymio website: `https://www.thymio.org` [Online; accessed 26-March-2023].

[16]Pheeno website: `https://discourse.ros.org/t/pheeno-a-low-cost-ros-compatible-swarm-robotic-platform/2698` [Online; accessed 26-March-2023].

[17]Spiderino website: `https://spiderino.nes.aau.at` [Online; accessed 26-March-2023].

[18]Zooids website: `https://github.com/ShapeLab/SwarmUI` [Online; accessed 26-March-2023].

[19]APIS website: `https://github.com/wvu-irl/reu-swarm-ros` (software only) [Online; accessed 26-March-2023].

deployment. Another wheeled robot is the Mechalino, that will be available open-source under Github[20]. Mechalino is a robot built from 3D-printed and commercial-off-the-shelf components. It is small and low-cost, but also very modular concept, allowing for extending the platform with extra sensors, actuators or computational capabilities if needed.

In the swarm logistics domain the authors in Jones et al. [186] built the DOT, a swarm robotic platform that is remotely accessible and comes with a complete toolchain to develop, simulate, and deploy code on the platform. In addition, to 5G and multiple cameras, it comes with a lifting platform on top. The WsBot [230] is another experimental platform for Industry 4.0 applications with a wireless charge system. It is manly designed with forklifts to achieve intelligent behaviors individually or in groups, where all agents are interconnected over Wi-Fi. The Milli-Robot [167] is equipped with magnetic hardware with up to 288 degrees of freedom and is proposed for general swarm robotic research activities. The authors introduce random dithering and argue to achieve a 100% success rate (i.e., no deadlocking). Another swarm robotic platform is the mROBerTO 2.0 [106] limited in size equipped with a locomotion mechanism that utilizes stepper motors to foucs on complex trajectories. This allows the robot to micro-step down to 1/32 of a full step.

The Droplet [208] is a swarm robotic platform for teaching and research. The spherical robot is able to organize and cluster into complex shapes with its neighbors. They use vibration motors for their locomotion. A specialty is that the robotic platforms charge and communicate via a powered floor that is equipped with alternating stripes of positive charge and ground. It is available as an open-source project[21]. The Swarm-bots [253, 149] can assemble themselves to different geometric 3D shapes. The robots are built by a number of simpler, insect-like robots that use relatively cheap components (the design is open-source[22]). These robots can perform self-assembly and self-organization to adapt to the current situation in their environment. This assembling capability allows the robots in swarm formation to transport objects that would be too heavy for an individual robot. The follow-up project of the Swarm-bots is the Swarmanoid project. This project counts as the very first attempt to study the integration of design, development, and control of a swarm of heterogeneous, open-source[23] swarm robotic platforms. The swarm in the Swarmanoid project presents three different types of autonomous robots, each equipped with various sensors: i) Eye-bots are UAVs that have the ability to attach to an indoor ceiling, ii) hand-bots are UGVs that are able to climb, and iii) foot-bots are UGVs that have a self-assembling capability [89]. The ARGroHBotS [277] is another swarm robotic platform where the user can adjust parts of the hardware (mainly

[20]Mechalino Github page: `https://github.com/smartgrids-aau/Mechalino_Arena` [Online; accessed 26-March-2023].

[21]Droplet website: `https://code.google.com/archive/p/cu-droplet/` [Online; accessed 26-March-2023].

[22]Swarm-bots website: `http://www.swarm-bots.org/` [Online; accessed 26-March-2023].

[23]Swarmanoid website: `http://www.swarmanoid.org/` [Online; accessed 26-March-2023].

3D printed pieces) to set it up for specific implementations. Additionally, the platforms allow remote access. The Termes robots [286] use modular blocks and create large structures without communication or GPS localization. They utilize the concept of stigmergy that is inspired by the way living termites build their nests. The Termes robots themselves have the abilities to climb and carry blocks to create these structures in unstructured, rough environments. Symbrion and Replicator [199] are two sibling projects, that develop autonomous swarm robotic platforms. They can be used as single entities or can be physically connected to form special patterns The main goal of these projects was to focus on the possibility to achieve evolvability of robot organisms. PolyBots [96] are self-reconfigurable robots. They have various types of locomotion options and object manipulation modules that are interchangeable to form a number of shapes, for example, an earthworm type to slither through obstacles, or a spider to stride over hilly terrain. These robots find application especially in unknown environments or multiple task challenges. Further modular robots that allow self-configuration with similar robotic technologies include M-TRAN [261], M-TRAN II [217] and M-TRAN III [218] (available as open-source[24] project), ATRON [46], CONRO [57], Sambot [403], Molecube [429], Puzzlebot [420], to name a few[25].

Finally, a remarkable new field of the application of swarm robotic platforms is the acoustic swarm [176] that creates speech zones representing a self-distributing wireless microphone array. In their experiments the authors showed that their swarm is able to localize and separate three to five concurrent speech sources without the usage of external infrastructure (such as a camera system).

6.4.2 Aerial

There are already multiple miniature UAVs available that form a good basis for research in aerial swarm robotic systems. A comprehensive overview of such small-scale UAVs can be found in Cai et al. [50] and Swetha et al. [373]. Off-the-shelf Micro Air Vehicles (MAVs) are available in a wide range of models and are widely used in the gaming industry, and other businesses for video and photography. Unfortunately, their flight controllers are typically closed so that we cannot develop and test custom algorithms (e.g., Qualcomm Flight Pro[26], DJI M100[27]). A UAV that is designed specifically for usage in swarms is the MAV presented by Roberts et al. [313]. These UAVs are equipped with three rate gyroscopes, three accelerometers, one ultrasonic sensor, and four IR sensors. It has been co-developed in the Swarmanoid project [89]. In the

[24]M-TRAN website: `https://www.wevolver.com/wevolver.staff/m-tran` [Online; accessed 26-March-2023].

[25]For a full list, the reader is referred to `https://en.wikipedia.org/wiki/Self-reconfiguring_modular_robot` [Online; accessed 26-March-2023].

[26]Qualcomm Flight Pro website: `https://www.lantronix.com/products/qualcomm-flight-pro-development-kit/` [Online; accessed 26-March-2023].

[27]DJI M100 website: `https://www.dji.com/at/matrice100/info\#specs` [Online; accessed 26-March-2023].

Distributed Flight Array [280] each UAV makes up a module of a larger array, but has only a single rotor. The individual modules are able to self-assemble to form a multi-rotor system, where all vehicles must cooperate constantly for a coordinated flight. To achieve this behavior, they exchange information among each other and, using the neighbor's information, they adjust local parameters. The Crazyflies, available both open-source and commercially[28], are used to realize a swarm of UAVs indoors. They are equipped with multiple sensors, e.g., accelerometer, gyroscope, magnetometer, and a high precision pressure sensor [293]. They have quite a low weight of 27 g that reduces danger for humans during the experiments. For the localization the Crazyflies need an external tracking system such as OptiTrack[29]. Another indoor swarm can be build with the FINken-III [158][30] and its predecessors. They are equipped with optical flow, IR distance, and a tower of four sonar ranging sensors.

6.4.3 Aquatic

In the CoCoRo (Collective Cognitive Robotics) project [330] a huge swarm of 41 heterogeneous UUVs has been developed. For the experiments they use three types of vehicles: A base station USV, an exploration UUV, and a UUV for relaying information between the base station and the explorers. Communication is achieved using sonar and electric fields. The primary envisioned applications include environmental monitoring, assessing water pollution and evaluating the effects of global warming. The UUV Monsun [278] uses two different types of communication: An acoustic underwater modem for information exchange and a camera to recognize and follow other swarm members. The CORATAM (Control of Aquatic Drones for Maritime Tasks) [62] project develops swarms of USVs for future environmental monitoring, sea life localization, and sea border patrolling. The USV platforms are available open-source[31] and execute swarm algorithms generated using evolutionary computation [94]. Another project is concentrating on the development of indoor USVs specifically the microUSV [146] which measures just 23 cm in length. They are still in the early stages and are currently capable of following waypoints. Nevertheless, indoor USVs for swarm robotics application are really necessary to perform behavioral deployment and research at low cost.

---

[28] bitcraze Crazyflies website: `https://www.bitcraze.io/crazyflie-2-1/` [Online; accessed 26-March-2023].

[29] Optitrack website: `https://optitrack.com/` [Online; accessed 26-March-2023].

[30] FINken website: `https://www.ci.ovgu.de/SwarmLab/Robots/FINkens.html` [Online; accessed 26-March-2023].

[31] CORATAM website: `http://biomachineslab.com/projects/control-of-aquatic-drones-for-maritime-tasks-coratam/` [Online; accessed 26-March-2023].

6.4.4 Outer Space

The NASA developed the Swarmies[32] for space exploration. The main goal is to collect material samples such as water, ice, or useful minerals on Mars. This application is referred to as in-situ resource utilization. Simultaneously, NASA launched a swarmathon[33] to motivate students in the development of swarm algorithms based on ant foraging. In the experiments 20 Swarmies can travel a linear distance of 42 km in 8 hours. This is the same distance that was covered by Mars rover Opportunity in 11 years. Another innovative project was accepted by the NASA Innovative Advanced Concepts program, namely the Marsbees [192]. Their objective is to enhance Mars exploration using flying swarm robotic platforms. They have the size of a bumblebee and self-explore the environment. The Mars rover Opportunity serves as base and charging station.

6.5 Project: How to Build your own Swarm Robot

6.5.1 A Robot with Legs: Spiderino

The Spiderino robot is an example of a legged robot that is aimed at swarm robotic experiments and educational purposes. Two major challenges for building a swarm robot, cost and size, are particularly difficult to master when it comes to legged robots. For instance, if a leg is equipped with two servo motors, one for lifting the leg and the other for moving it forward or backward, a robot with six legs would require several servo motors to operate and control. One possibility would be reducing the number of legs, but that would complicate the movement and possibly require extra sensors for the control system. For example, biped walkers require sensory systems to keep balance while moving over unknown ground. When aiming at a system which is statically stable at all stages of their stride, at least four legs are required. A quadruped following this principle would lift only one leg at a time and select the sequence of movements so that the center of gravity is always above the area spanned by the feet on the ground. Turtles are frequently mentioned as an example of a gait that maintains static stability. However, in reality, a turtle's stride can also include phases where the body is supported only by two diagonally opposite legs [6, 428]. For a sixed-legged robot, an effective gait involves lifting three legs at a time while maintaining stability with the other three legs. This approach allows for faster movement compared to the statically stable one-leg-at-a-time gait used by quadrupes. Therefore, the challenge remains: How can we design the mechanics in a cost-efficient and compact way.

[32] Swarmies website: `https://www.nasa.gov/content/meet-the-swarmies-robotics-answer-to-bugs` [Online; accessed 26-March-2023].

[33] Swarathon website: `https://swarmathon.cs.unm.edu/` [Online; accessed 26-March-2023].

Table 6.1: Classification of research platforms for swarm robotics (adapted from Schranz et al. [337]).

Environment	Application	Project/ Product Name	Robot Type	No. of Robots	Basic Swarm Behaviors	Availability
Terrestrial	Research and Education	Kilobots	UGV	1024 [316]	pattern formation, coordinated motion	open-source, commercial
		Jasmine		60 [198]	aggregation, collective exploration, coordinated motion, task allocation, collective perception, self-healing (partially), human-swarm interaction (Zooids, APIS)	open-source
		Alice		20 [132]		n.a.
		AMiR		6 [21]		
		Colias		14 [19]		open-source, commercial
		Mona		30 [17]		
		R-One		n.a.		n.a.
		Elisa-3		38 [137]		open-source, commercial
		Khepera IV		10 [289]		commercial
		GRITSbot		100 [287]		open-source
		E-Puck		16 [8]		open-source, commercial
		Xpuck		16 [187]		n.a.
		Thymio II		8 [387]		open-source, commercial
		Pheeno		4 [405]		open-source
		Spiderino		n.a.		
		I-Swarm		n.a.		n.a.
		Zooids		32 [223]		open-source
		APIS		6 [82]		n.a.
		Wanda		11 [200]		
		Mechalino		4		open-source
		DOTS		20 [186]		n.a.
		WsBot		2 [230]		open-source
		Milli-Robot		16 [167]		n.a.
		mROBerTo		n.a.		n.a.
		Droplet		n.a.	aggregation, self-assembly, object clustering and assembly, collective exploration, coordinated motion, collective transport (partially), collective perception	open-source
		Swarm-bot		35 [149]		
		Swarmanoid		n.a.		
		ARGroHBots		15 [277]		open-source
		Termes		5 [286]		n.a.
		Symbrion and Replicator		n.a.		
		PolyBot		32 [96]		
		M-Tran III		24 [218]		open-source
		ATRON		7 [46]		
		CONRO		8 [57]		n.a.
		Sambot		15 [407]		
		Molecube		8 [429]	pattern formation, self-assembly, self-reproduction	
		Puzzle Bot		9 [420]		open-source
		Acoustic Swarm		5 [176]	collective exploration, coordinated motion, collective perception	open-source
Aerial		MAV	UAV	n.a.	n.a.	n.a.
		Distributed Flight Array		9 [279]	self-assembly, coordinated motion	
		Crazyflie 2.1		49 [293]	aggregation, collective exploration, coordinated motion, collective localization, collective perception	open-source, commercial
		FINken-III		n.a.		n.a.
Aquatic	Environmental Monitoring	CoCoRo	UUV	41 [330]	aggregation, collective exploration, collective localization, task allocation	n.a.
		Monsun		n.a.		
		CORATAM	USV	12 [94]		open-source
		microUSV		4 [146]		open-source
Outer Space	Space Exploration	Swarmies	UGV	20[1]	collective exploration, collective localization	n.a.
		Marsbee	UAV	3 [192]	collective exploration, coordinated motion, task allocation	

Figure 6.6: Spiderino robot (left) and its components.

In the case of the Spiderino robot, the control system has been put atop an existing system that comes form a Hexbug Spider toy model (see Figure 6.6). The Hexbug Spider toy is a remote controlled plastic toy featuring six legs being controlled by two motors, one for moving the robot forward or backward and the other for turning the robot. A smart mechanical design enables the robot to move by lifting three legs at a time in the direction the robot's head is facing. The toy, being mass-produced, comes at a retail price between 15 and 30 €, making the system economically attractive even if the parts for the remote control are removed and not used in the robot.

To build your own Spiderino robot, follow these steps:

Electronics: The PCB file of Spiderino is available for download at the Spiderino website[35]. The main board is a two-layer, circular board of 80 mm diameter. The top layer contains most of the components, including

- Socket for Arduino Pro Mini.
- Connector for an ESP8266 WiFi Module.
- Socket for motor driver (POLOLU Motor DRV-DRV8835).
- Three-way power switch.
- Three-pin configuration jumper.
- 5 resistors and 2 LEDs, both in 1206 SMD format.

[35]Spiderino website: `https://spiderino.nes.aau.at/index.php/downloads/[Online; accessed26-March-2023]`.

Connecting the WiFi module is optional since the robot can also be programmed through a USB cable and is capable of operating without a WiFi connection.

The bottom layer of the PCB includes 6 connectors for sensor boards and connectors for the two DC motors.

The sensor board is a very small 10 by 10 mm 2-layer PCB that connects through four pins to the main board. Two pins are used for GND and VCC, while the other two pins contain a digital I/O and an analog input. The board can fit different sensors. The current sensor boards have an optical CNY70 distance sensor that points radially outward from the robot. Up to six sensor boards can be fit onto the robot, but it is also possible to have a configuration with fewer sensors.

Mechanical structure: In addition to the PCB, a Hexbug Spider toy robot and two 3D-printed adapters are required. The toy's main body has to be separated from its head containing the IR remote control interface since control will be implemented through the PCB. The main board is connected through screws to a 3D-printed adapter board that holds the battery and contains a cut-out for the motor cables. This part is glued to another conical adapter that fits the lower part of a Hexbug Spider toy robot, which contains the motors and provides a mechanism for coordinated leg actuation.

Programming: The robot's control program has to be uploaded to the Arduino Pro Mini. The Pro Mini can be programmed with Arduino IDE (available for Windows, Linux, and MacOS systems) in C or Atmel assembly language. The C-language API consists of functions for controlling the robots' motion and reading the sensors. Its implementation is documented in the technical documentation available for download at the Spiderino webpage[180].

6.5.2 A Wheeled Robot: Mona

Mona wheeled robot [17] is an open-source miniature robot that was mainly developed for long-term swarm robotics applications. It was first tested for a Perpetual Robot Swarm [24] system that aimed to develop a low-power robotic system with extremely long autonomy time, over weeks or months. Mona is an autonomous mobile robot with five proximity sensors, utilizes an AVR microcontroller as its main controller and can be programmed by popular Arduino IDE. The Mona robot has been commercialized[36] however it is still an open-source platform.

Two off-the-shelf DC motors with direct reduction gears were used as the main actuator. As shown in the architecture of the robot in Figure 6.7, a symmetrical differential driven configuration is used to control the robot's

[36]Mona robot commercial website: `https://ice9robotics.co.uk/` [Online; accessed 26-March-2023].

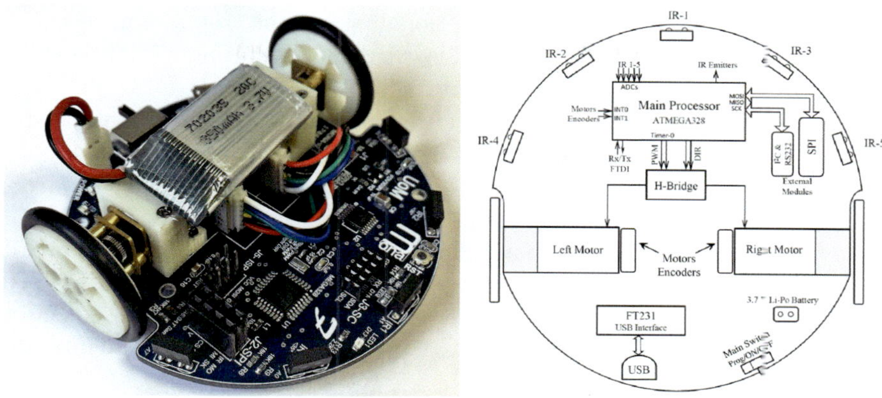

Figure 6.7: (left) Mona robot main platform. (right) The architecture of the basic platform [17].

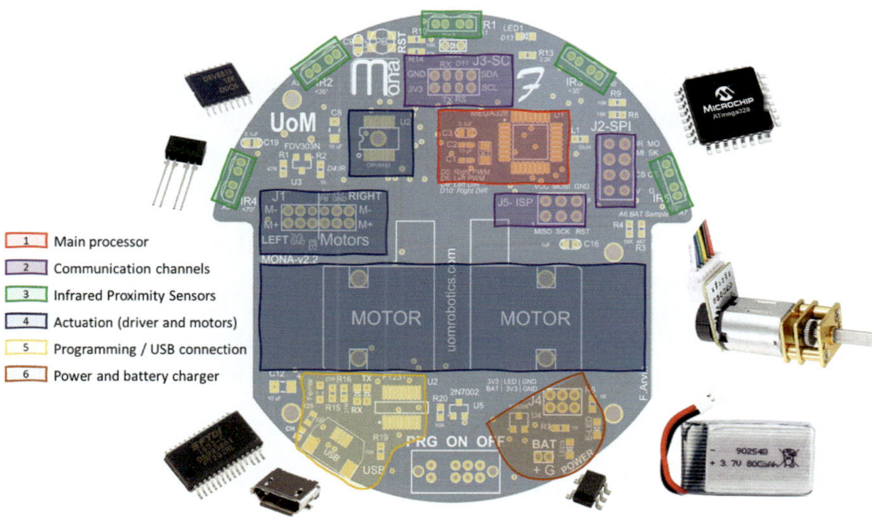

Figure 6.8: The Mona main-board and various sub-modules.

motion with two motors. The robot's wheelbase is 80 mm with 28 mm wheel diameters. The rotational speed of each motor is independently controlled through a pulse-width modulation channel of the controller, which is connected to a H-bridge motor driver.

To build your own Mona robot, one has to consider three crucial steps:

Electronics: The PCB file of the Mona is available online[37]. It is a two-layer PCB and most of the components are on the top board. The main board has six functions (as shown in Figure 6.8) including: (1) The main controller that is an ATMEGA328, (2) communication connectors supporting RS232, serial peripheral interface, I^2C (Inter-Integrated Circuit) and ISP, (3) IR proximity sensors, (4) motor driver and feedback link from motor encoders, (5) USB port and asynchronous serial data module used for programming the robot with Arduino, and (6) 3.7 V LiPo battery and the charging unit.

Mechanical structure: The main PCB also acts as the chassis of the robot, so motors, battery and all mechanical components sit on the main-board. This makes the assembly and controlling of the robot easier. In terms of mechanical structure, there are two parts—wheels and motor brackets—that must be 3D printed. The CAD design of the parts is also shared at the Mona GitHub.

Programming: The robot is compatible with most of the open-source programming platforms. Although the architecture of the robot enables programming Mona with any Arduino-based platforms via a USB cable, any programming language that is available for the AVR μCs, including C, C++, Java, Pascal, Basic, and Assembly can also be used for programming the robot. Hence, any software platform and IDE that support ISP can be used for programming the robot.

The latest version of the Mona robot was equipped with an ESP32 micro-controller providing additional functionality such as WiFi, Bluetooth and RGB LEDs. There are various available examples of robotic modules used for undergraduate engineering students. It includes a library (`Mona.h`), several examples for controlling motors and an autonomous obstacle avoidance scenario available at: `https://github.com/MonaRobot/Mona-Arduino`.

6.5.3 Another Wheeled Robot: Mechalino

The design of Mechalino is guided by the requirements for

- A low-cost robot (around 150 € per robot) so that a swarm of these robots is still affordable,
- Using wheels with good grip and stepper motors to have accurate movement control,
- Having a mostly cylindrical body in order to avoid getting stuck on edges,

[37]Mona Test Github: https://github.com/Farshad-Arvin/Mona-Platform/blob/master/Mona-Test.ino [Online; accessed 26-March-2023].

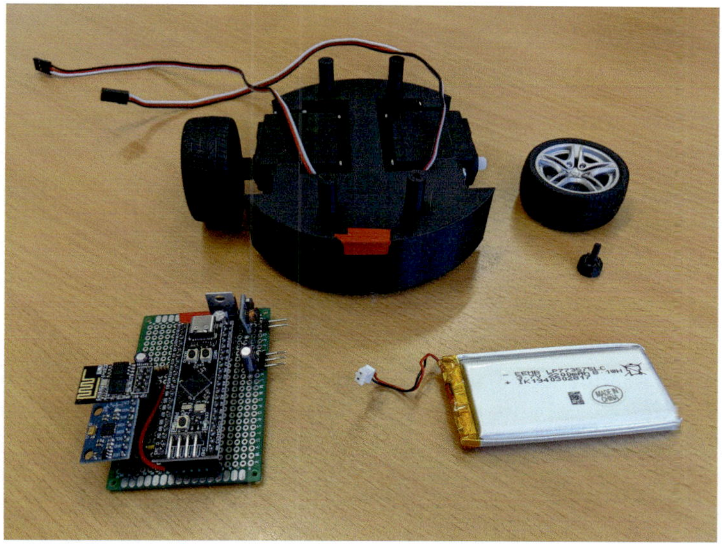

Figure 6.9: Mechalino components: Chassis, wheel, axle, mainboard, battery.

- Providing an area on top for adding a marker that can be detected by an overhead camera.

While still reasonably priced, Mechalino is designed to support ROS and provide the processing power for running more elaborated algorithms and evaluations onboard. It is equipped with a STM32F4 Microcontroller Unit (MCU). The software and hardware design is available under an open-source license[38]. For easy reproducibility of the hardware, the design aims to use standard components in combination with parts that are easily producible via 3D printing. Figure 6.9 shows the main components of a Mechalino robot. The assembled prototype is shown in Figure 6.10. The color markers on top allow for the accurate detection of the position and orientation of the robot from an overhead camera. The robot has a low center of gravity, facilitating precise movement. On the hand hand, the low ground clearance requires to operate the robot on a smooth surface. Mechalino is still under development and is being continuously improved.

To build your own Mechalino robot, follow these steps:

Electronics: The PCB file is available for download at the Mechalino webpage. The two-layer PCB fits the microcontroller, an ESP8266, a GY-521 gyroscope/acceleration sensor, two driving circuits for the left and the right motor and distance sensors.

[38]Mechalino website: `https://github.com/smartgrids-aau/Mechalino_Arena` [Online; accessed 26-March-2023].

Figure 6.10: Assembled Mechalino with ArUco marker on top.

Mechanical structure: The main body features a 3D-printed chassis with cut-outs for the motors and the battery. At the top, there are screw holes for attaching the main board, also allowing for a possible second-layer processing board. The wheels are pressureless standard rubber wheels from a model shop. A 3D-printed sleeve ensures a snug fit to the stepper motor axles.

Programming: The robot can be programmed in C++ and in ROS using MicroPython[39]. A hardware abstraction layer needs to be uploaded onto the STM32F4, which is available at the project page. Programming is done through a cable, while the ESP8266 is utilized for communication at runtime.

[39]MicroPython website: https://micropython.org/ [Online; accessed 24-December-2024].

Chapter 7

Open Challenges and Outlook

As applications become more complex, there are specific challenges that need to be addressed. Traditional technical solutions involve creating a final product that includes numerous components and sub-components. Each of these components has a significant size and complexity of its own and must interact with other components to provide their services in a specified manner. Even minor changes to a component's services or interface could potentially disrupt the entire system's functionality. Existing examples have shown potential for using swarm-based approaches in engineering in order to address issues that come with the growing complexity of systems via a self-organizing approach. In addition, technical applications become more complex due to an increasing number of networked components being involved on both sides, the environment and the technical solution. The previous chapters have introduced several examples showing how swarm-based approaches can be successfully employed in such an environment. However, these examples might just mark the beginning of using swarm systems as a new paradigm for designing and operating systems. The complexity of applications in terms of functionality, components, and dependencies among these components will rise further, which means that we can expect the existing challenges to grow, as well in their importance. To face these challenges, new methods, techniques, and approaches need to be developed. By successfully solving these challenges, integrating swarm-based approaches to overcome existing technological boundaries will be a promising direction to explore and increase the efficiency of technical applications. The following sections are adapted and extended from Schranz et al. [333].

7.1 Challenges in Swarms of CPSs

As already discussed in Section 1.2, the modeling, design and control challenges for CPS swarms become even more difficult as the autonomy of individual components must be combined with explicit attention to interdependencies and coordination, interoperability, distributed control and emergence of behaviors. Swarm intelligence can bring potential advantages in terms of adaptability, robustness, and scalability to systems. However, several research challenges remain, especially when it comes to particular applications. For example, applying swarm intelligence to the physical domain, aka CPSs. In this domain, the limitations of embedded systems have to be considered, as well as an incomplete knowledge of the state of the physical world. Addressing these challenges not only helps to overcome current limitations but also paves the way for future applications of swarm systems in various fields.

7.1.1 Challenges Deriving CPS and Swarm Properties

Today's technical systems are usually predictable and controllable. For example, entering a target temperature in an oven causes the expected reaction, namely that the oven heats up to the set temperature and maintains this temperature. In swarm systems, these properties are often not directly given, since such a system is characterized by local rules, interactions and thus the emergence of the global system goal from the bottom up. A control input for a swarm system is, therefore neither easy to realize, nor is it guaranteed that the result of an input is easily predictable. In the design of CPS systems, the question of predictions, for instance, about the next system state, movement, or information, has a direct impact on the ability to give the system certain configurations, such as manipulating parameters, controlling individual movements, or assigning specific tasks. Thus, in addition to the issues of predictability and controllability, any parameter setting, even if only slightly adjusted, can significantly affect the behavior and dynamics of an intelligent swarm system. In particular, for a CPS swarm system, predicting such effects is difficult because modeling of the physical parts of the system is usually subject to uncertainty [333].

In recent years, there have been several research works addressing these topics, in particular related to network science [135] and the observability of complex systems [232]. Many systems are characterized by a number of parameters, numerical and structural, whose effects on the resulting dynamics in a complex system can often be extreme, leading to phase transitions [360, 76, 42], self-organized criticality [402] and deterministic chaotic behaviors [369, 423].

In CPS swarms, these parameter effects come from several sources that are typically hard to model, including environmental effects such as weather, earthquakes, the (partially) unpredictable behavior of humans, or any other random disturbance coming from the real world. The difficulty lies in creating

precise models for timely adaptation to exhibit the desired robustness to the dynamics of the environment [34]. Therefore, swarm-intelligent systems may instead find limited use in life-critical applications such as the control of an airplane—a clear definition of the possible behaviors of the system cannot be provided in a CPS swarm system.

Literature reports several works on the analysis of formal models for optimization or decision-making scenarios. They mostly use the Ant Colony Optimization (ACO) and the PSO models in combination with optimization problems and analyze stability and convergence properties. The analyzed scenarios often rely on constrained models, such as simplified algorithms or additional assumptions [65]. Examples include using a random dynamical system, for PSO [105], Lyapunov stability analysis [188], or deriving graphical parameters from dynamical systems theory [383]). In many practical applications, ACO and PSO models become too complex for complete theoretical analysis [333]. Typically, performance analysis is done via empirical procedures, possibly restricted on the assumption of defined theoretical properties [333]. Examples of such an approach involve using a specifically designed objective function for convergence analysis of a PSO approach [66], reviewing self-adaptive PSO algorithms related to stable points in parameter selection [154], parameter optimization studies [422], running numerical tests and results using test matrices associating multi-target tracking and measurement [44], and applying experimental analysis of how particular design choices affect the quality and the shape of the Pareto front approximations generated in multi-objective ACO [235].

In contrast to swarm intelligence systems in general, a formal model of CPS swarms to evaluate predictability and controllability is much more sensitive. In CPS swarms, we do not only consider the swarm itself, but also the different performance objectives, metrics, constraints, requirements, and disturbances coming from the real-world problem scenario. In addition to that, we have the cyber (behaviors and interactions) and physical (hardware-related) characteristics of each individual CPS. This means that each variation of the CPS, its behaviors, the scenario they operate impose a new set of configurations and thus, challenges in practice. This makes it difficult to abstract general properties and reuse results across different scenarios [333].

Nevertheless, with extensive mathematical abstractions it is possible to identify and analyze a basic set of swarm behaviors, including, foraging [58, 136, 95, 162], coverage [424, 338, 237, 69], aggregation and pattern formation [134, 13, 136], and cooperative tracking [424, 142, 340] (see Section 1.1.1 and Section 4.1 for more details); an extensive, and well-discussed list of swarm behaviors can be found in Brambilla et al. [45], Bayindir [31], and Schranz et al. [332].

7.1.2 Challenges Designing the Local Rules

The design of swarm intelligence often follows a bottom-up approach, where the desired global behavior is implicitly determined by defining local

interaction rules. This requires solving an inverse problem using heuristics to guide the design of the local rules. In swarm intelligence scenarios that are not tied to a physical scenario, for instance, generic optimization applications of swarm intelligence [301], the definition of heuristics that lead to a desired global behavior is easy because they are predetermined by the optimization goal.

However, in embodied swarm applications, such as CPS swarms, this task is much more complex due to the multi-layered, uncertain and hardly predictable structure of the environment. Therefore, most of the work in this direction is devoted to the automatic synthesis of behavioral and interaction rules for swarm robots. In Fehervari and Elmenreich [111], artificial evolutionary techniques (provided by FREVO [359]; see Section 4.5 for more details) are used to generate neural controllers for homogeneous robot soccer teams. In Gomes et al. [144] they apply novelty seeking to generate neural controllers for homogeneous robot swarms. Ferrante et al. [115] generate controllers for robots by combining the use of an evolutionary approach with a formal language. In Lopes et al. [234], supervisory control theory, a formal language based on a discrete-event representation of the system, is used to automatically generate controllers with built-in proofs and wider reusability in different robotic systems. Similarly, in Francesca et al. [124], the automatic design of robot swarm control software is achieved by generating a probabilistic finite state machine resulting from an optimization process that maximizes a task-specific objective function. Furthermore, in Tuci et al. [384], different evolutionary approaches are used to generate task assignment mechanisms that efficiently adapt robot behavior to the environment.

In general, these approaches, even if exemplified in a specific application case, can be extended to other domains. For example, adapting an algorithm from autonomous ground robots to autonomous aerial vehicles should be possible after considering additional constraints (aerial vehicles cannot easily bump into each other without damage) and degrees of freedom (height as a third dimension). All of these works, among others, mark valuable contributions to the automatic generation of rules that can be reused or expanded for the more general case of CPSs. However, it is important to acknowledge that these approaches have only been validated in relatively simple scenarios and have been applied exclusively to homogeneous swarms. As a result, these methods need to be better equipped to handle the complexities of heterogeneous swarms in CPSs.

Furthermore, these methods face challenges when dealing with large and complex search spaces, leading to what is commonly known as the dimensionality curse. Consequently, current automatic generation systems may not be feasible for systematic use when synthesizing complex behaviors required by CPS swarms. It is worth noting that a convergence between model-checking methods [64] for general CPSs and automatic rule synthesis and verification in physical swarms will be necessary to provide guarantees in both the design and control stages of CPS swarms. Such a combination would

ensure a systematic and reliable approach to address the challenges associated with CPS swarm systems.

7.1.3 Real-World Deployment Challenges

To date, most of the work implemented and tested on physical swarms has focused on swarm robotic research platforms. These works typically involve small, simple robots with basic sensors and limited data exchange capabilities. Although they exhibit interesting self-organizing behaviors, they serve mainly as test demonstrations and are difficult to scale for real-world applications. Their reliability, predictability, and efficiency need to be reevaluated when the same approach is to be applied to complex tasks in realistic environments. In contrast, contemporary technological advances have produced impressive individual CPS and robotic systems. Examples include self-driving cars developed by companies such as Uber and Google, remarkably stable quadrupedal robots such as Boston Dynamics' Spot, and advanced humanoid robots such as Boston Dynamics' Atlas, Hanson Robotics' Sophia, and Honda's Asimo. These systems are highly complex mechanically, algorithmically, and behaviorally. Compared to these cutting-edge examples, the tasks performed by current swarms of robots seem rudimentary. The question arises whether swarm robotics or specifically CPS swarms, can progress from being study objects to becoming systems producing real-world systems that can reliably perform useful tasks.

This transition may require a move away from overly simplified robot models and controls toward designs that achieve a better balance between simplicity and the ability to perform complex tasks effectively and reliably. It could entail transitioning from minimal resource usage and restricted information exchange to a more extensive utilization of sensor data and enhanced information sharing. Intel's Shooting Star [310] drone swarm provides an illustrative example in this regard. During the opening ceremony of the 2018 Winter Olympics, Intel orchestrated an impressive aerial show with 1218 drones. Although labeled as a swarm, the system is not really swarm intelligent; it operates as a partially distributed system with a central controller. Individual drones follow pre-computed trajectory scripts and rely on a precise external positioning system for navigation. The main idea is that to effectively perform real-world tasks, it might be essential to abandon overly restrictive assumptions and adopt a pragmatic, balanced approach by combining bottom-up swarm intelligence design with top-down methods whenever feasible.

Another significant challenge associated with the implementation of real swarm systems is the maintenance of such systems throughout their operational lifetime. As discussed in previous research [97], although they are swarm systems, complex technical systems that must operate reliably over a long period of time will require maintenance to ensure uninterrupted service. It remains an open question whether maintaining a swarm system with its unique characteristics will be more accessible or more challenging than traditionally

designed technical applications. In essence, maintaining a system composed of numerous autonomous components will be a challenging task. However, multiple components within the system can also provide opportunities for internal monitoring and troubleshooting. A relevant example is given in Christensen et al. [63], where failed robots are promptly detected and isolated from the swarm of robots, demonstrating the potential for addressing problems within the system itself.

7.1.4 How to Address the Challenges

In order to enhance the potential for formal analysis and promote the systematic reuse of results in swarm intelligence, it is advantageous to position swarm intelligence within the wider framework of complex systems and network science. This approach allows the adoption of sophisticated mathematical and modeling tools commonly used in the study of complex systems, which can help characterize properties such as evolution over time, stability, and structural/topological aspects. The focus on structural and topological properties has recently gained significant attention, as information flow plays a crucial role in self-organization and emergence [143, 79].

In addition, it is vital to identify the relationship between swarm intelligence and the fields of game theory and multi-agent systems [351]. Although this connection has yet to be adequately explored, exploiting insights from game theory and adopting mechanism design approaches can provide viable alternatives for generating interaction rules and analyzing the formal properties of swarm intelligence systems. Concepts such as equilibrium in game theory and interaction rule design through mechanism design [272, 41] present promising avenues for automatic rule generation. Given the current limitations of CPS swarms, it is expected that the study of physical swarms should align more closely with the multi-agent systems field while focusing on self-organizing behavior and scalability. The distinction between swarms and decentralized multi-agent systems is not always clearly defined. A recent survey of multi-robot systems [70] shows that the properties required for distributed control algorithms include locality of sensing and communications, scalability, security, contingency, and task-oriented considerations, similar to those addressed in research on swarm robotics [31]. This suggests that the differences between the two concepts are often semantic rather than substantive, with considerable overlap between them (for more details refer to the discussion in Section 1.1.3).

7.2 Future Trends and Directions

As pointed out in the introduction, there is a growing presence of complex dynamic systems that are increasingly collective and connected. These systems may be cyber or physical or, more generally, both, constituting a cyber-physical system. Given the nature and characteristics of these systems, swarm

intelligence will play an essential role in addressing the upcoming challenges. Given the vast solution possibilities and diverse challenges, new swarm models may be needed, and new fields of application will open up. In the following, we discuss where these new models might come from for inspiration and in what kinds of future applications swarm intelligence will prove to be a relevant tool. Finally, we outline general open research topics on CPS swarms.

7.2.1 Future Inspirations

Until today, most swarm intelligence models and algorithms have been inspired by social interactions within animal groups and animal societies [281]. Such observations are expected to remain a constant source of inspiration for new swarm intelligence models. However, we can also expand the range of potential inspiration models to include organic, inorganic, and social systems with similar fundamental properties. We can identify essential microscopic behaviors and local interaction rules by reducing these models to the key functional elements required in most swarm intelligence models. These microscopic interactions generate self-organizing behavior as an emergent feature at the macroscopic level.

One challenge is to build and extract abstract models from natural sources of inspiration. This includes exploring decentralized and time-efficient techniques from the specific scenarios studied in different research disciplines. Secondly, these models need to be used to build new models and address the problem complexity of real-world CPS applications.

The following sections illustrate specific areas that can provide new ideas for swarm intelligence model development. In some cases, a model has already been used as inspiration for swarm intelligence development. However, their potential to inspire swarm intelligence systems has yet to be fully exploited, so even an example known to have inspired swarm intelligence algorithms could be used to make new and potentially (more) disruptive contributions.

Biology

Animal societies have been the primary source of inspiration in swarm intelligence. Researchers and swarm intelligence practitioners have drawn inspiration primarily from cooperative animal societies, largely ignoring predator-prey systems and other conflicting scenarios [281]. However, to address cyber threats and the interaction among multiple, potentially competing CPSs, it is necessary to study non-collaborative systems. Research in population dynamics [174], game theory [351], and evolutionary game theory [322] has provided applicable models and ideas for the development of novel swarm intelligence systems. A promising approach involves the integration of autonomous robots into existing animal societies Successful integration has been demonstrated with honeybees, fish, cockroaches, and cows [251, 326, 399]. This emerging field offers the potential to create a bio-hybrid system, or social cyborg, that combines the unique capabilities

of robots and animals in a mutually beneficial and symbiotic way. *(Programmable) Bacteria* serve as an inspiration for current and future research in various fields, including social intelligence. Their foraging [281] and collective decision-making abilities have allowed them to thrive and spread in challenging environments, making them a subject of interest for researchers seeking to develop new methods for solving complex problems [350]. In addition to their natural abilities, bacteria can also be programmed through genetic engineering techniques, providing opportunities for precise control and manipulation of their behavior [184]. For example, bacteria can be programmed to produce specific proteins or enzymes or respond to certain environmental signals. One exciting application of programmed bacteria is in the field of medicine. Recent research has shown that a swarm of magnetically-guided bacteria can be used to deliver drugs directly to tumors within the body, potentially improving the efficacy of cancer treatments [113]. This approach offers a promising alternative to traditional drug delivery methods, which can have harmful side effects and may not effectively target cancer cells.

Chemistry and Physics

Molecular networks have been studied across various fields [191], particularly for modeling complex diseases such as cancer and schizophrenia [147]. In these networks, nodes represent molecules like genes, RNA, and proteins; the edges represent their relationships. These network structures could inspire the design of complex interactions between agents in a swarm, as they exhibit outstanding properties such as scalability and resilience. For instance, the physiological interaction network of a single cell comprises thousands of chemical reactions that alter the concentration levels of hundreds or thousands of chemical components. Despite the enormous complexity and size of these networks, they exhibit quick and flexible dynamics while staying within precisely controlled bounds, thereby ensuring the survival of the organism. The properties of molecular and bacterial networks, characterized by their flexibility, robustness, scalability, and emergent complexity resulting from simple interactions, share remarkable similarities with the desired characteristics of swarm intelligence. Conducting laboratory experiments under controlled conditions provides an opportunity to study the characteristics of these networks. Because of the relatively small scale, a study can be conducted, for instance, on a few microliters of fluid in a cuvette or a small agar plate on a laboratory desk. Thus, these networks present ideal subjects for systematic exploration to provide valuable inspiration for developing future swarm intelligence algorithms.

Nanonetworks refer to a set of nanomachines that are interconnected and have various capabilities such as computation, data storage, sensing, and actuation. The research on nanonetworks has focused primarily on their non-traditional communication types, such as electromagnetic or molecular communication [3]. In particular, molecular communication between cells

using synthesis, transformation, emission, propagation, and reception of molecules has the potential to inspire swarm intelligence [4]. One of the main advantages of molecular communication is its ability to function in environments that are not conducive to electromagnetic signals, such as inside the human body. This makes it a promising avenue for designing swarm intelligence systems operating in similar environments, such as healthcare or environmental monitoring applications. Moreover, molecular communication offers a high degree of security and privacy due to the localized nature of the communication. Nanonetworks have already been successfully employed in biomedical applications such as drug delivery, in-vivo sensing, and monitoring glucose levels for diabetes patients. They have also been used in environmental applications such as pollution monitoring. Therefore, the study of nanonetworks and their communication mechanisms could serve as a rich source of inspiration for the design of future swarm intelligence algorithms. By leveraging the capabilities of nanonetworks, swarm intelligence systems could potentially achieve higher efficiency, scalability, and robustness in a wide range of applications.

Human Cognition

Humans also employ self-organizing strategies to interact with each other to solve various tasks [213, 215, 109], such as the collective patterns observed in crowd dynamics. Experimental studies have been conducted to identify corresponding behavioral rules, such as investigations of pedestrian behavior in single avoidance tasks [258]. In a study by Tavakoli et al. [377], they observed the coordination abilities of humans with limited perception in an environmentally distributed environment. This first experiment aimed to learn human-inspired behavior before formal strategies, potentially adaptable to CPS swarms, were developed. Surprisingly, compared to the multitude of algorithms derived from (eu)social animals, very little research has focused on extracting behaviors from human groups or societies to convert them into swarm intelligence algorithms [282]. This is unexpected since humans, being the creators of advanced and complex societies and cultures, are a readily available source of inspiration, easily accessible through natural language communication.

Interdisciplinary Approaches

The swarm-like concept of *active matter* consists of numerous individual agents following simple rules, resulting in collective behaviors and movements. Active matter combines representations from various disciplines, such as biology, physics, computer graphics, and robotics. A well-studied model in this domain is the self-propelled particle model proposed by Vicsek et al. [394] in 1995. This model has been extended to simulate various realistic systems, including self-replicating morphogenesis, in which swarms form structures that create new similar structures [328].

The researchers described collective motion in active solids and crystals, applying elasticity-based mechanisms to achieve self-organization [117, 116]. These models can be adapted to study the collective movement of CPS swarms, such as pattern formation or morphogenesis. They offer robustness to heterogeneity, which can be derived analytically and is particularly suitable for CPSs.

7.2.2 Promising Future Applications

Although the current implementation of swarm intelligence in real-world scenarios remains limited, we recognize the vast potential for its application in a wide range of practical situations. As Bonabeau and Meyer have pointed out, the possibilities for swarm intelligence application are limited only by our imagination [40]. This statement strongly supports our hypothesis, stated in the introduction, regarding the ubiquitous presence of swarm intelligence in meeting the challenges of an increasingly interconnected world.

Swarm intelligence has already been successful in various complex optimization scenarios in the form of algorithms such as ACO [84] and PSO [197], where swarm members encode individual solution candidates. However, physically embedded swarm algorithms have seen limited applications so far, with most engineering solutions following more traditional top-down or centralized approaches.

We expect that swarm intelligence will increasingly find applications in real-world scenarios where a top-down approach lacks sufficient information, a centralized approach is computationally infeasible, or real-time constraints impede the ability to find optimal solutions. These situations often arise in complex CPSs or scenarios without existing infrastructure, where the need is to provide ad-hoc working solutions without the ability to first establish an infrastructure framework (e.g., in disaster scenarios [78]). Furthermore, we foresee significant potential for swarm applications when transitioning from well-defined environments such as planned factories or warehouses to more complex environments such as traffic systems, urban environments, social networks, extreme/hostile environments, or uncharted territories.

Looking ahead, we consider the next frontier for swarm intelligence in CPS swarms. In this regard, we highlight specific application scenarios that are currently (partially) unattainable but have the potential to take full advantage of swarm characteristics. Much research already exists, many of which await system integration to meet future application requirements. In the following examples, we describe the intended application of CPS swarms, outline their swarm characteristics, and identify areas of research that require further investigation to achieve the desired applications, where the term "CPS swarm" can be understood as an umbrella term.

Autonomous Driving and Smart Traffic

Autonomous driving is a prominent and much discussed topic, presenting various computational challenges in dynamic environments requiring real-time processing capabilities. This is where the concept of swarm intelligence becomes relevant: Thousands of cars, each with different levels of autonomy, as well as the road infrastructure, collaborate to find common solutions. Applying a swarm model to autonomous cars and infrastructure in an intelligent traffic management system offers several benefits, including reduced traffic congestion, coordinated creation of emergency lanes, improved traffic flow, and reduced carbon emissions. One potential application is swarm or fleet navigation, incorporating information from other swarm members, infrastructure, and the dynamic environment.

Modeling networked autonomous cars and smart infrastructure as a swarm is both feasible and beneficial for several reasons:

1. Each CPS can retrieve and evaluate information locally, following local rules.

2. Local information can be exchanged between CPSs to form a global or swarm-level perspective.

3. The infrastructure is relatively known in advance or subject to slight dynamics, such as accidents or construction sites.

A central traffic management system would have difficulty coordinating and managing each CPS participating in the traffic. At the same time, individual CPSs may cooperate locally, potentially leading to emergent behaviors.

While transportation logistics already makes extensive use of ant- and bee-inspired algorithms for route optimization [255, 421, 306, 426, 346], the concept of smart traffic remains relatively unexplored. However, the media frequently talks about swarms of autonomous cars and smart infrastructure [126, 339, 168], indicating the growing interest in such applications. However, implementing these swarms requires careful consideration, especially in terms of management. Questions arise, such as: What motivates CPSs to form a swarm? What common features, properties, or sensor data should CPSs share? How do CPSs join or leave a swarm? What decision-making processes should prioritize local driver goals, swarm goals, or global smart traffic goals? What is the optimal number of CPSs needed to form and benefit from a swarm? Integrating the data obtained from the infrastructure into the autonomous car swarm could further improve the decision-making process to achieve the above goals.

The technology to realize CPS swarms in smart traffic partially exists today. Local retrieval and evaluation of sensor data in cars is already common practice, and efforts are underway to develop standardized communication protocols for vehicle-to-vehicle and vehicle-to-infrastructure communication, collectively called Vehicle to Everything (V2X) [59]. V2X involves communication between cars and smart infrastructure integrating

physical elements like streets, cameras, traffic lights with digital infrastructure like sensors and communication networks. Communication technologies such as 5G [12] and Ultra-Wideband [152] offer high levels of security, low latency, and high data rates, enabling self-location and real-time data exchange between CPSs. These advances enable a swarm of autonomous cars to share and react to real-time local information, adapting to dynamic situations. For example, in Ulbrich et al. [388], the authors describe a swarm behavior module that selects and summarizes sensor data and performs trajectory planning for an autonomous car. In Li et al. [227], they develop swarm-intelligence-based local rules to enable the efficient operation of interconnected autonomous cars on arterials.

Honda's SAFE SWARM^{TM} concept is an example of improving traffic by using information from vehicles in front, including onboard sensors, systems, and V2X [256] communication. Implementing an efficient and effective intelligent traffic system would likely require a complete redesign of the existing architecture for real-time data collection, processing, and analysis. In addition to the swarm system, a central Traffic Management System (TMS) could be implemented to fulfill several supporting roles. First, the TMS would function as a data submission system, ensuring that local event information is available to all other cars so they can plan their routes accordingly. Second, the TMS would provide an overview of global traffic situations by integrating and merging information about individual road swarms. This would provide a global understanding of traffic conditions and facilitate more effective decision-making. The data exchange process between cars and the TMS can be further enhanced by incorporating information from the intelligent road infrastructure. By integrating data from sensors embedded in the infrastructure, such as traffic lights or cameras, the overall system can gather more complete and accurate information about road conditions. In summary, the development of an intelligent traffic system would involve not only the implementation of a swarm system for autonomous cars but also the implementation of a central TMS. This would require a new architecture for real-time data collection, processing, and analysis. By incorporating information from both the vehicles and the smart road infrastructure, the system can achieve higher levels of efficiency and effectiveness in managing traffic flows.

Emergency Response

The presence of professional first responder teams in emergency situations is critical to saving lives and restoring essential services. Traditional first responder teams consist of police, military, firefighters, and rescue units from civil protection agencies. However, the use of CPSs swarms can greatly enhance and facilitate emergency response activities. These CPS swarms, including different types of robots such as all-terrain or flying robots [78, 325], can work collaboratively with human teams, creating a heterogeneous, multi-human CPS swarm. These mixed teams can effectively handle accidents at

industrial or power plants, as well as environmental disasters such as wildfires, floods, storms and earthquakes. CPS agents within the swarm have the potential to navigate destroyed or inhospitable environments more efficiently and safely than humans. They can collect multi-sensor data, coordinate with human team members, focus on areas of interest like potential survivors, remove obstacles, restore structures and save lives. In essence, a CPS swarm can serve as a dynamic and distributed augmented sensor-actuator system, providing invaluable support to human first responders in effectively and efficiently conducting their missions.

Joint teaming with CPS swarms is feasible under several conditions: (i) Each individual swarm member can operate under local rules and is independent of the fixed infrastructure. (ii) The infrastructure is often disrupted, with blocked or unstable paths, making it a dynamic environment. (iii) Individual information can be exchanged between swarm members, other swarms and local control stations. (iv) A dynamic ad-hoc communication infrastructure is established to replace the pre-existing infrastructure that may no longer be available. This allows for intra- and inter-swarm communication, as well as communication between the swarm and control centers. (v) High-level decision making can be done in the control centers, balancing local autonomy with system-level decision making. (vi) Swarm members and human teammates can interact and exchange information transparently at the local level. (vii) Cooperative and collaborative behaviors are encouraged at the individual level to enable effective behaviors to emerge at the global level.

To effectively deploy mixed teams (heterogeneous swarms) in real-world scenarios, substantial progress is still needed, particularly in autonomous decision-making, data exchange and fusion, and operation in complex and unexplored environments. Research efforts are underway to address these challenges in heterogeneous swarms, such as those involving drones and ground rovers working together to provide first aid or guide victims to emergency exits. For example, Saez-Pons et al. [319] developed an autonomous robot swarm for SAR applications. They used social potential fields to generate formations and navigate while maintaining them. Although several other projects in this area [80, 238, 372], none have yet produced applicable results for real-world missions. The design and deployment of a heterogeneous swarm for such missions involve various open questions, including determining the number, type, and initial location of robots [71].

In the context of human interaction in mixed teams, it is essential to design intuitive and seamless methods for multimodal dialogue and data exchange between humans and CPS. This requires developing user-friendly interfaces and controls [148, 314, 216, 92]. Infrastructure, including communication networks, is often destroyed in emergency response applications, such as post-earthquake scenarios. Therefore, new relaying concepts are needed to enable the CPS swarm to create its own communication network. For example, Hauert et al. [156] developed a swarm of autonomous micro aerial vehicles capable of deploying and managing an ad-hoc Wi-Fi network. Furthermore, the highly coordinated and collaborative nature of the swarm members is

crucial in the disrupted environment. This coordination can be achieved by exchanging status updates or, if bandwidth permits, by sharing complete images and detected points of interest. It is important to develop compact image processing algorithms that align with local processing capabilities.

Environmental Monitoring

Monitoring Earth's threatened ecosystems is an emerging application for large-scale autonomous CPS swarms. These swarms can function as active sensor networks that adapt based on previously collected and assessed habitat data. This enables monitoring points of interest with greater accuracy, achieved by increasing the density of CPSs or improving the frequency of measurements at specific locations.

Supporting environmental monitoring through CPS swarms is feasible due to several factors: (i) Each individual swarm member operates under local rules, facilitating autonomous behavior. (ii) Information exchange occurs between CPSs to determine new collective behaviors. (iii) The natural environment being monitored is generally unknown and subject to dynamic and unpredictable events. (iv) Environment monitoring is intended to avoid disruption from external intruders, emphasizing the need for a high degree of autonomy. (v) Cooperation among individual swarm members leads to emergent behavior, enabling achieving monitoring objectives.

Swarms are well suited to operate in unexpected environments and can compensate for the loss or failure of swarm members. Therefore, this technology is an ideal candidate for long-term autonomous operation [83] during unpredictable events in unknown environments. For environmental monitoring, lightweight and efficient communication protocols between CPS [331] are especially appropriate. Excessive communication can disrupt the natural communication network of the ecosystem. Minimizing CPS communication and actions preserves the undisturbed nature of the environment, making it easier to analyze authentic natural behavior and processes. Therefore, adopting a principle such as "act only when necessary, but do so intelligently and gently" aligns with the behavioral optimization found in natural organisms. In addition, light and efficient communication protocols contribute to energy conservation. Since these swarms are typically deployed for months or even years to monitor ecosystems, smart energy consumption must align with the observation time.

A CPS swarm designed for environmental monitoring can consist of actively moving agents, allowing them to perform a crucial task at the termination of the monitoring process: Self-removal of the swarm from ecosystems in an environmentally friendly and sustainable manner, ensuring no harm is done. Several research projects are known to have successfully used real robots for environmental monitoring, including subCULTron [380], CORATAM [94], and CoCoRo [330].

Electric Energy Grids

Electric energy grids are highly complex interconnected systems and, therefore, prime candidates for swarm-based solutions. Venayagamoorthy [393] identifies several application areas where computational intelligence can be used in electric power grids, such as algorithms for energy and power flow management, voltage and reactive power control, dynamic load forecasting, and vehicle-to-grid integration. Although swarm algorithms are not explicitly mentioned as examples, many of these coordination tasks might benefit from swarm intelligence approaches. In addition to using swarm-based optimization algorithms such as PSO, there are approaches that directly model a portion of the electric power grid as a swarm. For example, Elmenreich et al. [102] present an example of coordinating homes in a neighborhood using a swarm-like approach to demand management. This approach uses geographic proximity to ensure fairness among consumers in the neighborhood. Steber et al. [364] describe a virtual mass storage system consisting of distributed battery energy storage units installed in homes with rooftop photovoltaic systems. Another example is the SmartGRID concept presented in Huang et al. [173], which proposes a decentralized and interoperable grid scheduling framework using a swarm-intelligent approach. Ramachandran et al. [304] proposes a hybrid swarm-immune-based auction system to coordinate generation and consumption in a smart microgrid. Black electrification using microgrids is another application that integrates small off-grid systems into an interconnected microgrid, as discussed in Kirchhoff et al. [205]. This approach, known as black-based electrification, is particularly relevant for rural, underdeveloped areas [220, 260].

The consideration of using CPS swarms to model electric power networks is compelling for several reasons: (i) The increasing use of distributed renewable energy sources is leading to more decentralized systems. (ii) Renewable energy sources such as solar PV and wind power cannot easily adjust their output to meet higher demand, making centralized control challenging. (iii) Electrification of transportation leads to increased electricity consumption by numerous individual consumers. These factors require a transition from centralized control of a few manageable power plants to a self-organized, swarm-like system in which generators, storage, and consumers interconnect in a distributed manner.

Practical deployment of CPS swarm models in real power grids requires significant research progress. Given the critical nature of electric power grids as essential infrastructure, validation and testing of such systems prior to deployment is a frequent requirement and an open question with respect to the current state of the art. Without validation and testing before deployment, implementing new control strategies, even in a small part of the system, can cause oscillations and affect the stability of the overall system [141]. Addressing this challenge may involve simulation at a granularity and accuracy so that the reality gap is neglicible [257] or the application of CPS swarm models within smart microgrids [357]. Smart microgrids provide

an environment large enough to benefit from a swarm-based approach while being isolated from the national grid, reducing potential risks.

Space Missions

Space missions offer tempting opportunities to use swarm-based CPS approaches. One compelling application is the inspection of communication satellites, where swarms can be used to assess hull damage and ensure optimal performance. In addition, coordinated swarms can be instrumental in addressing the space debris issue by performing disposal operations. Another interesting concept involves using a cluster of reflective spheres to measure the gravitational force of asteroids in our solar system. These missions involve exploring water, raw materials, and potential life on planets and exoplanets, which serve as potential precursors to human colonization efforts.

Key features required for these missions include low-cost performance and fault tolerance, which can be achieved by using large CPS swarms that operate autonomously in unknown environments [368]. Establishing reliable communication channels to relay the data collected by the swarm back to Earth is crucial. For example, the Marsbees concept [219] envisages using the Mars rover as a base and recharging station for these autonomous CPS swarms. The Japan Aerospace Exploration Agency is taking this idea further by envisaging swarms of autonomous machines to prepare the ground, excavate and build facilities for astronauts on the Moon and Mars [419]. The first real-world tests were related to networking and communication in different spheres including Nodes [51], Proba-3 [291], and KickSat [204].

The feasibility of using CPS swarms for such visionary applications stems from several factors: (i) Each individual CPS operates based on local rules, which allows for autonomous behavior. (ii) Local information is exchanged between the CPS and stations on Earth, potentially using relay stations in planetary orbits. (iii) These deployment scenarios involve dynamic and unknown environments with no prior knowledge of the infrastructure. (iv) Centralized control is impractical due to the significant time delays between swarm operation and human intervention. (v) Successful achievement of mission objectives requires effective cooperation between individual swarm members.

Currently, the above missions still rely heavily on costly ground control for their operations. However, the ultimate vision is establishing autonomous satellite formations, as human intervention causes communication delays and associated costs. Formations are closely linked to the motion of spacecraft, which is influenced by factors such as gravity, solar radiation pressure, and atmospheric drag, causing deviations from the motion patterns observed on Earth [77]. The development of motion models becomes crucial for the guidance, navigation, and control of these formations. When direct exploration of planetary surfaces becomes feasible, autonomous swarms can be used. However, challenges remain related to transportation to the planet, launch procedures, navigation, ground surveying, and data retrieval. Factors such as

planet-specific temperatures, ground conditions, and gravity must be carefully considered during these operations.

Medical Applications

The field of medicine has shown a growing interest in using swarms to solve complex problems, such as cancer treatment. In healthcare, nanoparticles are important because they can leak out of blood vessels and go to their target sites. Nanoparticles play a key role in healthcare as they can penetrate blood vessels and target specific sites in the body. Although these particles are too small to be individually programmed, they can be prepared for swarm-based applications by changes to their coating, charge, or size. Suppose a group of nanoparticles exhibits swarming behavior. In that case, they have the potential to navigate to target cancer cells, carry a healing coating, be activated by external stimuli, and effectively destroy target tissues. Unlike other drugs that are dispersed throughout tissues, a swarm of nanoparticles has greater intelligence and can target diseased tissue more precisely.

Several factors support the use of nanoparticle swarms in medical applications: (i) Each individual nanoparticle operates based on local rules, which allows for autonomous behavior. (ii) Local information is exchanged between nanoparticles and possibly with an external monitoring station outside the body. (iii) In such scenarios, there is no existing infrastructure. (iv) The environment is a human or animal body, which is partly unknown and highly dynamic. (v) Achieving the desired goals requires cooperation between individual nanoparticles, leading to emergent behavior.

However, some fundamental questions remain unanswered when it comes to designing a swarm of nanoparticles for cancer therapy. For example, how to establish communication between individual swarm members, between swarm members and the environment, and between the swarm and the outside world? In addition, the implementation of swarm characteristics related to complexity, local intelligence, and local processing poses challenges. The swarm system must be simple, reconfigurable, cost-effective, scalable, and verifiable. Addressing these issues could provide insights into implementing more complex tasks such as optimization, computation, decision-making, construction, self-assembly, and collective movement within the swarm.

This vision of cancer treatment based on nanoparticle swarms is still in its early stages. However, it raises the question of whether other medical applications could be realized by injecting the swarm into the human body. Further research is needed to fully understand the possibilities and implications of this exciting medical frontier.

Human Networks

Human organizations and social networks, powered by Web 2.0 (defined as the transition from static websites to dynamic user-generated content and social media), represent collective systems. The swarm-like nature of social

networks becomes apparent when a large number of individuals interact and communicate locally, shaping global behavioral patterns such as online trends and the popularization of products or ideas.

In this context, swarm intelligence methodologies can be applied to study such systems from a swarm point of view. Human networks communicating over the Internet come with the advantage to provdie vast amount of available online data sources. Analytical prediction and evaluation through simulation can help to predict and potentially control the behavior of other complex systems. Furthermore, swarm intelligence could find a direct practical application within human networks in two main ways: As an integral part of the network itself and as a physical service that can be used, traded, and negotiated by the network. We already use different types of digital assistants like Alexa [299], Siri [16], Cortana [166]) in our daily lives, which can be seen as some of the most advanced examples of CPSs in the real world. They communicate with people via text or speech, have full connectivity, are often mobile, perform computations to facilitate precise interactions with people, and exhibit proactive behavior such as suggesting or reminding tasks. With continuous advances, these digital assistants are becoming increasingly autonomous and capable of decision-making, adaptation, and lifelong learning. In the future, we can envisage the delegation of many tasks to these digital partners, allowing them to work on our behalf 24/7. When we reach this stage, the digital 'society of us' will parallel our physical society, among other things, performing information searches, making connections, and engaging in commerce.

The CPS swarm approach provides a framework for managing large-scale systems consisting of billions of agents. This approach offers several advantages: (i) Agents in a swarm act autonomously and are guided by a set of behavioral rules that reflect the personality traits and preferences of their human counterparts. (ii) Agents cooperate locally based on the concepts of a dynamic social neighborhood, while exchanging and sharing global information. (iii) The environments in which agents operate, including the emerging Web 3.0 and the physical world, which are highly dynamic, difficult to predict, and inherently parallel and distributed. (iv) Digital assistants, a form of CPS swarm, are designed to provide services based on gathered information. In a swarm of digital assistants, there is a constant exchange of information, searching, negotiation and trading between members. (v) Each agent in the swarm is inherently selfish and represents the preferences of its human double. However, with appropriate rules of interaction, socially aware patterns can emerge globally.

There are many practical applications for CPS swarms of digital assistants. For example, with appropriate consent and privacy protection, the continuous exchange of information between digital assistants can enable the tracking of physical locations. This capability could play a key role in scenarios such as epidemic tracking, where potential exposures can be identified and containment measures automatically implemented. In emergency situations, swarm assistants can take advantage of real-time information sharing to guide

the safe escape of individuals in a crowd, allowing for the emergence of a collective escape plan. Another possible application is coming from the circulation of fake news on social networks that presents a serious problem. Cooperation between different web miners working for different entities is often limited, which hinders the collection of comprehensive information and effective inference. Using a collaborative CPS swarm, a large amount of information can be collected in parallel, shared among members, and used for coordinated analysis and inference, allowing for more effective detection of fake news. Looking to the future, CPS swarms of digital assistants could work together to find, negotiate and trade with other CPS swarms operating directly in the physical world. This multi-level interaction between CPS swarms opens up new possibilities for service provision and the resolution of complex tasks. In addition, sensing as a service is a growing business approach in IoT, where users pay for data collected by specific sensors. This concept can be extended to swarms of CPSs, allowing them to offer sensing and actuation services in different locations. Members of the digital swarm can then find and negotiate the use and cost of available CPS resources for specific tasks, sharing experiences and collaborating with each other. This creates a complex society of CPS swarms interacting at multiple levels. Blockchain technology, known for its decentralized and distributed nature, will play a fundamental role in enabling secure transactions and building trust between swarm members.

In summary, CPS swarm approaches provide a powerful framework for managing large-scale systems, and digital assistants serve as conspicuous examples of CPS swarms. The potential applications of CPS swarms are diverse, ranging from tracking epidemics and responding to emergencies, to combating fake news and creating complex societies of interacting swarms.

7.2.3 Research Challenges

As swarm intelligence is still a relatively young topic, several crucial open research questions remain. In the previous subsections, we discussed domain-specific topics related to specific innovative applications. However, it is essential to note that many other application-specific research questions are not mentioned here. These issues are highly dependent on the specific context of the application. Additionally, there are general themes in swarm systems that are independent of the domain or application. We will use the example of CPS swarms to illustrate these themes. A swarm of CPSs faces several requirements that need to be addressed. These include designing, programming, and implementing highly distributed and connected digital technologies embedded in a wide range of devices. In addition, designing increasingly autonomous physical systems with diverse dynamics is challenging while meeting several critical constraints. In addition, addressing systems of systems with a high degree of autonomy is key to ensure scalability, adaptability, robustness, complexity management, safety, and security, and establishing trust between people and swarms. In order to provide an overview of open research topics in swarm systems, we group general topics into the

following categories. It is important to note that these categories address only a subset of the general topics, as a discussion of all the underlying details would be beyond the scope of this book.

Modeling

Creating models from natural or other sources of inspiration is an important topic in the research community. As mentioned in Section 7.2.1, many untapped sources of inspiration can still be explored. This research area also includes fine-tuning application-specific parameters for existing algorithms. In addition, algorithmic topics cover scalability, degree of control, degree of convergence, and the delicate balance between exploitation and exploration [418]. Furthermore, combining different sources of inspiration can be used to solve complex modern problems with self-organizing capabilities. The key idea is to adapt the application-specific swarm intelligence algorithms to fit the underlying problem definition.

Design

As swarms of CPS continue to proliferate in different application contexts and experience increasing acceptance, the challenge of designing systems that can efficiently achieve predefined goals while remaining flexible, reliable, and adaptable to changing conditions is becoming increasingly daunting.

The design of the next generation CPS must address multiple broad challenges, as described by Isaac et al. [67]. These challenges include: (i) Integration of complex, heterogeneous large systems, (ii) Interaction between people and systems, (iii) Dealing with uncertainty, (iv) Measuring and validating system performance, (v) Enriching systems with learning capabilities and (vi) System design.

From a fundamental point of view, individual theories provide formal descriptions of different aspects of CPS design, covering physical, technical, organizational, and human-system interaction aspects. However, these disciplines are not fully integrated into a single system theory. In other words, while methodologies, representations, and tools exist to address specific aspects of CPS design, there is still an open challenge in providing support for the complete design life cycle. In addition, a formal design methodology must include a process that iteratively revisits and refines micro-level behavior and bridges the gap between local and global behavior. Although Brambilla et al. [45] describe behavior-based design and automated design methods for CPS swarms, there is currently no implemented toolchain that can map these design processes. One possible approach to address this gap is the CPSwarm workbench [25], which integrates several tools to guide CPS swarm designers through the entire lifecycle, covering the design, optimization, simulation, and deployment phases.

Validation and Verification

Verification and validation in CPS swarms cover a wide range of systems and systems of systems, including hardware, software, information, processes, personnel, and facilities. It is crucial to explore robust methodologies for validating swarm behavior in CPSs. Whenever possible, it is important to define detailed standardized KPIs, scenarios, test areas, and benchmarks to ensure the reusability of measurement methodologies. The results obtained from verification and validation activities should feed back into the design specifications, model definitions, and validation.

Human in the Loop

Enabling human interactions with swarm systems [269, 210] offers many advantages, as CPSs can leverage humans' cognitive and sensory-motor capabilities. In contrast, humans can use CPSs in closed-loop interactions as augmented external sensor-actuator systems. Nevertheless, the design and implementation of such mixed-initiative systems [182] present scientific and technological challenges. These challenges encompass various aspects, including the physical performance of two-way interactions (between single CPSs or swarms of CPSs and humans), the presentation of the complex state of the distributed CPS to humans, and the dissemination of information or commands from humans to humans and CPS swarms. Considerable research has been devoted to the development of bi-directional interaction and dialogue interfaces and modalities, exploring the use of various multimodal interfaces [148, 49, 270, 185] for proximal interaction with swarms.

However, allowing direct interaction of human operators within a swarm system introduces new potential risks. The presence of humans can negatively impact the security, safety, stability, and reliability of the system response if they act irresponsibly or maliciously. These aspects have not received much attention to date but will become major concerns in the near future, especially with respect to security and privacy at all levels.

Bibliography

[1] Russ Abbott. Emergence explained: Abstractions: Getting epiphenomena to do real work. *Complexity*, 12(1):13–26, October 2006.

[2] Faisul Arif Ahmad, Abd Rahman Ramli, Khairulmizam Samsudin, and Shaiful Jahari Hashim. Optimization of power utilization in multimobile robot foraging behavior inspired by honeybees system. *The Scientific World Journal*, 2014, 2014.

[3] Ian F. Akyildiz, Josep M. Jornet, and Massimiliano Pierobon. Nanonetworks: A new frontier in communications. *Communications of the ACM*, 54(11):84–89, 2011.

[4] Ian F. Akyildiz, Massimiliano Pierobon, Sasitharan Balasubramaniam, and Yevgeni Koucheryavy. The internet of bio-nano things. *IEEE Communications Magazine*, 53(3):32–40, 2015.

[5] Sergey Alatartsev, Vera Mersheeva, Marcus Augustine, and Frank Ortmeier. On optimizing a sequence of robotic tasks. In *Proceedings of the IEEE/RSJ International Conference on Intelligent Robots and Systems*, pages 217–223, 2013.

[6] R. McNeill Alexander. Gaits of mammals and turtles. *Journal of the Robotics Society of Japan*, 11(3):314–319, 1993.

[7] Antonio Luca Alfeo, Eduardo Castello Ferrer, Yago Lizarribar Carrillo, Arnaud Grignard, Luis Alonso Pastor, Dylan T. Sleeper, Mario G.C.A. Cimino, Bruno Lepri, Gigliola Vaglini, Kent Larson, Marco Dorigo, and Alex Sandy Pentland. Urban swarms: A new approach for autonomous waste management. *IEEE International Conference on Robotics and Automation*, 2019-May:4233–4240, 2019.

[8] Muhanad H Mohammed Alkilabi, Aparajit Narayan, and Elio Tuci. Cooperative Object Transport With a Swarm of e-puck Robots: Robustness and Scalability of Evolved Collective Strategies. *Swarm Intelligence*, 11(3-4):185–209, 2017.

[9] Michael Allwright, Navneet Bhalla, Haitham El-faham, Anthony Antoun, Carlo Pinciroli, and Marco Dorigo. Srocs: Leveraging stigmergy

on a multi-robot construction platform for unknown environments. In *Swarm Intelligence: 9th International Conference, ANTS 2014, Brussels, Belgium, September 10-12, 2014. Proceedings 9*, pages 158–169. Springer, 2014.

[10] Lee Altenberg. Advances in genetic programming. In Kenneth E. Kinnear, Jr., editor, *The Evolution of Evolvability in Genetic Programming*, chapter 3, pages 47–74. MIT Press, Cambridge, MA, USA, 1994.

[11] Arash Sadeghi Amjadi, Mohsen Raoufi, Ali Emre Turgut, George Broughton, Tomáš Krajník, and Farshad Arvin. Cooperative pollution source exploration and cleanup with a bio-inspired swarm robot aggregation. In *International Conference on Collaborative Computing: Networking, Applications and Worksharing*, pages 469–481. Springer, 2020.

[12] Jeffrey G Andrews, Stefano Buzzi, Wan Choi, Stephen V Hanly, Angel Lozano, Anthony CK Soong, and Jianzhong Charlie Zhang. What will 5g be? *IEEE Journal on Selected Areas in Communications*, 32(6):1065–1082, 2014.

[13] Gianluca Antonelli, Filippo Arrichiello, and Stefano Chiaverini. Flocking for multi-robot systems via the null-space-based behavioral control. *Swarm Intelligence*, 4(1):1–37, 2009.

[14] Miguel Aranda, Gonzalo López-Nicolás, Carlos Sagüés, and Youcef Mezouar. Formation control of mobile robots using multiple aerial cameras. *IEEE Transactions on Robotics*, 31(4):1064–1071, 2015.

[15] Vincent Arnould, Laurent Balmelli, Ian Bailey, James Baker, Cory Bialowas, Conrad Bock, Carolyn Boettcher, Roger Burkhart, Murray Cantor, Bruce Douglass, Harald Eisenmann, Anders Ek, Brenda Ellis, Marilyn Escue, Sanford Friedenthal, Eran Gery, Hal Hamilton, Dwayne Hardy, James Hummel, Cris Kobryn, Michael Latta, John Low, Robert Long, Kumar Marimuthu, Alan Moore, Véronique Normand, Salah Obeid, Eldad Palachi, David Price, Bran Selic, Chris Sibbald, Joseph Skipper, Rick Steiner, Robert Thompson, Jim U'Ren, Tim Weilkiens, Thomas Weigert, and Brian Willard. Systems modeling language (SysML) version 1.5. Standard, Object Management Group (OMG), May 2017.

[16] Jacob Aron. How innovative is Apple's new voice assistant, Siri? *New Scientist*, 212(2836):24, 2011.

[17] Farshad Arvin, Jose Espinosa, Benjamin Bird, Andrew West, Simon Watson, and Barry Lennox. Mona: An Affordable Open-source Mobile Robot for Education and Research. *Journal of Intelligent & Robotic Systems*, 94(3-4):761–775, 2019.

[18] Farshad Arvin, Tomáš Krajník, Ali Emre Turgut, and Shigang Yue. COSΦ: artificial pheromone system for robotic swarms research. In *IEEE/RSJ international conference on intelligent robots and systems (IROS)*, pages 407–412, 2015.

[19] Farshad Arvin, John Murray, Chun Zhang, and Shigang Yue. Colias: An Autonomous Micro Robot for Swarm Robotic Applications. *International Journal of Advanced Robotic Systems*, 11(7):113, 2014.

[20] Farshad Arvin, Khairulmizam Samsudin, and Abdul Rahman Ramli. Development of a Miniature Robot for SwarmRobotic Application. *International Journal of Computer and Electrical Engineering*, 1(4):436, 2009.

[21] Farshad Arvin, Khairulmizam Samsudin, Abdul Rahman Ramli, and Masoud Bekravi. Imitation of Honeybee Aggregation with Collective Behavior of Swarm Robots. *International Journal of Computational Intelligence Systems*, 4(4):739–748, 2011.

[22] Farshad Arvin, Ali Emre Turgut, Farhad Bazyari, Kutluk Bilge Arikan, Nicola Bellotto, and Shigang Yue. Cue-based aggregation with a mobile robot swarm: A novel fuzzy-based method. *Adaptive Behavior*, 22(3):189–206, 2014.

[23] Farshad Arvin, Ali Emre Turgut, Tomáš Krajník, Salar Rahimi, Ilkin Ege Okay, Shigang Yue, Simon Watson, and Barry Lennox. ϕclust: Pheromone-based aggregation for robotic swarms. In *2018 IEEE/RSJ International Conference on Intelligent Robots and Systems (IROS)*, pages 4288–4294. IEEE, 2018.

[24] Farshad Arvin, Simon Watson, Ali Emre Turgut, Jose Espinosa, Tomáš Krajník, and Barry Lennox. Perpetual Robot Swarm: Long-term Autonomy of Mobile Robots Using on-the-fly Inductive Charging. *Journal of Intelligent & Robotic Systems*, pages 1–18, 2017.

[25] Alessandra Bagnato, Regina Krisztina Bíró, Dario Bonino, Claudio Pastrone, Wilfried Elmenreich, René Reiners, Melanie Schranz, and Edin Arnautovic. Designing swarms of cyber-physical systems: The H2020 CPSwarm project. In *Proceedings of the Computing Frontiers Conference*, pages 305–312, 2017.

[26] Mazen Bahaidarah, Fatemeh Rekabi Bana, Ali Emre Turgut, Ognjen Marjanovic, and Farshad Arvin. Optimization of a self-organized collective motion in a robotic swarm. In *International Conference on Swarm Intelligence*, pages 341–349. Springer, 2022.

[27] Mazen Bahaidarah, Ognjen Marjanovic, Fatemeh Rekabi-Bana, and Farshad Arvin. An optimised robot swarm flocking with genetic algorithm. In *Proceedings of the IEEE International Conference on Mechatronics and Automation*, pages 1823–1828, 2023.

[28] Mazen Bahaidarah, Fatemeh Rekabi-Bana, Ognjen Marjanovic, and Farshad Arvin. Swarm flocking using optimisation for a self-organised collective motion. *Swarm and Evolutionary Computation*, page 101491, 2024.

[29] Zhe Ban, Junyan Hu, Barry Lennox, and Farshad Arvin. Self-organised collision-free flocking mechanism in heterogeneous robot swarms. *Mobile Networks and Applications*, pages 1–11, 2021.

[30] Cyrill Baumann and Alcherio Martinoli. A modular functional framework for the design and evaluation of multi-robot navigation. *Robotics and Autonomous Systems*, 144:103849, 2021.

[31] Levent Bayindir. A review of swarm robotics tasks. *Neurocomputing*, 172:292–321, 2016.

[32] Levent Bayindir and Erol Şahin. A review of studies in swarm robotics. *Turkish Journal of Electrical Engineering & Computer Sciences*, 15(2):115–147, 2007.

[33] Gerardo Beni and Jing Wang. Swarm intelligence in cellular robotic systems. In *Proceedings of the NATO Advanced Workshop on Robots and Biological Systems*, pages 703–712, 1989.

[34] Amel Bennaceur, Carlo Ghezzi, Kenji Tei, Timo Kehrer, Danny Weyns, Radu Calinescu, Schahram Dustdar, Zhenjiang Hu, Shinichi Honiden, Fuyuki Ishikawa, Zhi Jin, Jeffrey Kramer, Marin Litoiu, Michele Loreti, Gabriel Moreno, Hausi Muller, Laura Nenzi, Bashar Nuseibeh, Liliana Pasquale, Wolfgang Reisig, Heinz Schmidt, Christos Tsigkanos, and Haiyan Zhao. Modelling and analysing resilient cyber-physical systems. In *Proceedings of the International Symposium on Software Engineering for Adaptive and Self-Managing Systems*, pages 70–76. IEEE, 2019.

[35] S Binitha, S Siva Sathya, et al. A Survey of Bio Inspired Optimization Algorithms. *International Journal of Soft Computing and Engineering*, 2(2):137–151, 2012.

[36] Christian Blum and Xiaodong Li. Swarm Intelligence in Optimization. In Christian Blum and Daniel Merkle, editors, *Swarm Intelligence: Introduction and Applications*. Springer, 2008.

[37] Michael Bodi, Ronald Thenius, Martina Szopek, Thomas Schmickl, and Karl Crailsheim. Interaction of robot swarms using the honeybee-inspired control algorithm beeclust. *Mathematical and Computer Modelling of Dynamical Systems*, 18(1):87–100, 2012.

[38] Eric Bonabeau. Editor's introduction: stigmergy. *Artificial Life*, 5(2):95–96, 1999.

[39] Eric Bonabeau, Marco Dorigo, and Guy Theraulaz. *Swarm Intelligence: From Natural to Artificial Systems.* Oxford university press, 1999.

[40] Eric Bonabeau and Christopher Meyer. Swarm intelligence: A whole new way to think about business. *Harvard Business Review*, 79(5):106–115, 2001.

[41] Tilman Börgers. *An Introduction to the Theory of Mechanism Design.* Oxford University Press, 2015.

[42] Terry Bossomaier, Lionel Barnett, and Michael Harré. Information and phase transitions in socio-economic systems. *Complex Adaptive Systems Modeling*, 1(1):9, 2013.

[43] Laszlo Böszörmenyi, Manfred del Fabro, Marian Kogler, Mathias Lux, O. Marques, and Anita Sobe. Innovative Directions in Self-organized Distributed Multimedia Systems. *Multimedia Tools and Applications, Springer*, 51(2):525–553, 2011.

[44] Ali Önder Bozdoğan, Asım Egemen Yilmaz, and Murat Efa. Performance analysis of swarm optimization approaches for the generalized assignment problem in multi-target tracking applications. *Turkish Journal of Electrical Engineering and Computer Sciences*, 18(6):1059–1078, 2010.

[45] Manuele Brambilla, Eliseo Ferrante, Mauro Birattari, and Marco Dorigo. Swarm robotics: A review from the swarm engineering perspective. *Swarm Intelligence*, 7(1):1–41, 2013.

[46] David Brandt, David Johan Christensen, and Henrik Hautop Lund. ATRON robots: Versatility from Self-reconfigurable Modules. In *Proceedings of the International Conference on Mechatronics and Automation*, pages 26–32, 2007.

[47] Kyle Brown, Oriana Peltzer, Martin A Sehr, Mac Schwager, and Mykel J Kochenderfer. Optimal sequential task assignment and path finding for multi-agent robotic assembly planning. In *2020 IEEE International Conference on Robotics and Automation (ICRA)*, pages 441–447. IEEE, 2020.

[48] Arne Brutschy, Giovanni Pini, Carlo Pinciroli, Mauro Birattari, and Marco Dorigo. Self-organized task allocation to sequentially interdependent tasks in swarm robotics. *Autonomous agents and multi-agent systems*, 28(1):101–125, 2014.

[49] Jonathan Cacace, Riccardo Caccavale, Alberto Finzi, and Vincenzo Lippiello. Attentional multimodal interface for multi-drone search in the alps. In *Proceedings of the IEEE International Conference on Systems, Man, and Cybernetics*, pages 1178–1183, 2016.

[50] Guowei Cai, Jorge Dias, and Lakmal Seneviratne. A Survey of Small-Scale Unmanned Aerial Vehicles: Recent Advances and Future Development Trends. *Unmanned Systems*, 2(2):175–199, 2014.

[51] Sonja Caldwell. Nodes – network & operation demonstration satellite. `https://www.nasa.gov/centers/ames/engineering/projects/nodes.html`, 2015. [Online; accessed 9-July-2019].

[52] Scott Camazine. *Self-organization in biological systems.* Princeton University Press, 2003.

[53] Scott Camazine, Nigel R. Franks, James Sneyd, Eric Bonabeau, Jean-Louis Deneubourg, and Guy Theraula. *Self-Organization in Biological Systems.* Princeton University Press, 2001.

[54] Alexandre Campo, Álvaro Gutiérrez, Shervin Nouyan, Carlo Pinciroli, Valentin Longchamp, Simon Garnier, and Marco Dorigo. Artificial pheromone for path selection by a foraging swarm of robots. *Biological Cybernetics*, 103(5):339–352, 2010.

[55] Gilles Caprari, Patrick Balmer, Ralph Piguet, and Roland Siegwart. The Autonomous Micro Robot "Alice": A Platform for Scientific and Commercial Applications. In *Proceedings of the International Symposium on Micromechatronics and Human Science.-Creation of New Industry-(Cat. No. 98TH8388)*, pages 231–235. IEEE, 1998.

[56] Stefano Carpin, Mike Lewis, Jijun Wang, Stephen Balakirsky, and Chris Scrapper. Usarsim: a robot simulator for research and education. In *Proceedings 2007 IEEE International Conference on Robotics and Automation*, pages 1400–1405. IEEE, 2007.

[57] Andres Castano, Alberto Behar, and Peter M Will. The Conro Modules for Reconfigurable Robots. *IEEE/ASME transactions on mechatronics*, 7(4):403–409, 2002.

[58] Eduardo Castello, Tomoyuki Yamamoto, Fabio Dalla Libera, Wenguo Liu, Alan F. T. Winfield, Yutaka Nakamura, and Hiroshi Ishiguro. Adaptive foraging for simulated and real robotic swarms: The dynamical response threshold approach. *Swarm Intelligence*, 10(1):1–31, 2016.

[59] Shanzhi Chen, Jinling Hu, Yan Shi, Ying Peng, Jiayi Fang, Rui Zhao, and Li Zhao. Vehicle-to-everything (v2x) services supported by lte-based systems and 5g. *IEEE Communications Standards Magazine*, 1(2):70–76, 2017.

[60] Ahmad Reza Cheraghi, Karol Actun, Sahdia Shahzad, and Kalman Graffi. Swarm-sim: A 2d & 3d simulation core for swarm agents. In *2020 3rd International Conference on Intelligent Robotic and Control Engineering (IRCE)*, pages 1–10. IEEE, 2020.

[61] Chafika Chettaoui, Franck Delaplace, Pierre Lescanne, Mun'delanji Vestergaard, and René Vestergaard. Rewriting game theory as a foundation for state-based models of gene regulation. In *International Conference on Computational Methods in Systems Biology*, pages 257–270. Springer, 2006.

[62] Anders Lyhne Christensen, Sancho Oliveira, Octavian Postolache, Maria João de Oliveira, Susana Sargento, Pedro Santana, Luís Nunes, Fernando J. Velez, Pedro Sebastião, Vasco Costa, Miguel Duarte, Jorge C. Gomes, Tiago Rodrigues, and Fernando Silva. Design of Communication and Control for Swarms of Aquatic Surface Drones. In *Proceedings of the International Conference on Agents and Artificial Intelligence*, pages 548–555, 2015.

[63] Anders Lyhne Christensen, Rehan O'Grady, and Marco Dorigo. From fireflies to fault-tolerant swarms of robots. *Transactions on Evolutionary Computation*, 13(4):754–766, August 2009.

[64] Edmund M. Clarke, Thomas A. Henzinger, Helmut Veith, and Roderick Bloem, editors. *Handbook of model checking.* Springer, 2016.

[65] Christopher W. Cleghorn. *Particle Swarm Optimization: Empirical and Theoretical Stability Analysis.* PhD thesis, Department of Computer Science, University of Pretoria, South-Africa, 2017.

[66] Christopher W. Cleghorn and Andries P. Engelbrecht. Particle swarm variants: Standardized convergence analysis. *Swarm Intelligence*, 9(2–3):177–203, 2015.

[67] Isaac Cohen, David Corman, Jim Davis, Himanshu Khurana, Pieter J. Mosterman, and Stormo Lonny Prasad Venkatesh. Strategic opportunities for 21st century cyber-physical systems. Technical report, NSF, Steering Committee for Foundations in Innovation for Cyber-Physical Systems, 2012.

[68] Steve Cook, Conrad Bock, Pete Rivett, Tom Rutt, Ed Seidewitz, Bran Selic, and Doug Tolbert. Unified modeling language (UML) version 2.5.1. Standard, Object Management Group (OMG), December 2017.

[69] Nikolaus Correll and Alcherio Martinoli. Robust distributed coverage using a swarm of miniature robots. In *Proceedings of the IEEE International Conference on Robotics and Automation*, pages 379–384, 2007.

[70] Jorge Cortés and Magnus Egerstedt. Coordinated control of multi-robot systems: A survey. *SICE Journal of Control, Measurement, and System Integration*, 10(6):495–503, 2017.

[71] Micael S. Couceiro, Carlos M. Figueiredo, David Portugal, Rui P. Rocha, and Nuno M. F. Ferreira. Initial deployment of a robotic team - a hierarchical approach under communication constraints verified on low-cost platforms. In *Proceedings of the IEEE/RSJ International Conference on Intelligent Robots and Systems*, pages 4614–4619, 2012.

[72] Bruno Courcelle. Graph rewriting: An algebraic and logic approach. In *Formal Models and Semantics*, pages 193–242. Elsevier, 1990.

[73] Iain D Couzin, Jens Krause, Richard James, Graeme D Ruxton, and Nigel R Franks. Collective memory and spatial sorting in animal groups. *Journal of theoretical biology*, 218(1):1–11, 2002.

[74] Felipe Cucker and Cristián Huepe. Flocking with informed agents. *Mathematics in Action*, 1(1):1–25, 2008.

[75] Renan da Silva Tchilian, Ubirajara Franco Moreno, and Mariana Netto. Assisted teleoperation for a human-swarm interaction system. In *3rd IFAC Workshop on Cyber-Physical & Human Systems CPHS*, pages 602–607, 2020.

[76] Christopher Dabrowski. Catastrophic event phenomena in communication networks: A survey. *Computer Science Review*, 18:10–45, 2015.

[77] Simone D'Amico, Marco Pavone, Shailendhar Saraf, Abdulaziz Alhussien, Turki Al-Saud, Sasha Buchman, Robert Byer, and Charbel Farhat. Miniaturized autonomous distributed space system for future science and exploration. In *Proceedings of the International Workshop on Satellite Constellations and Formation Flying*, pages 1–20, 2015.

[78] Kai Daniel, Bjoern Dusza, Andreas Lewandowski, and Christian Wietfeld. Airshield: A system-of-systems muav remote sensing architecture for disaster response. In *Proceedings of the 3rd Annual IEEE Systems Conference*, pages 196–200. IEEE, 2009.

[79] Christian Darabos, Mario Giacobini, and Marco Tomassini. Performance and robustness of cellular automata computation on irregular networks. *Advances in Complex Systems*, 10(supp01):85–110, 2007.

[80] Geert De Cubber, Daniela Doroftei, Daniel Serrano, Keshav Chintamani, Rui Sabino, and Stephane Ourevitch. The EU-ICARUS project: Developing assistive robotic tools for search and rescue operations. In *Proceedings of the IEEE International Symposium on Safety, Security, and Rescue Robotics*, pages 1–4, 2013.

[81] Kevin DeMarco, Eric Squires, Michael Day, and Charles Pippin. Simulating collaborative robots in a massive multi-agent game environment (scrimmage). In *Distributed Autonomous Robotic Systems*, pages 283–297. Springer, 2019.

[82] Neel Dhanaraj, Nathan Hewitt, Casey Edmonds-Estes, Rachel Jarman, J. Seo, Henry Gunner, Alexandra Hatfield, Tucker Johnson, Lunet Yifru, Julietta Maffeo, Guilherme Pereira, Jason Gross, and Yu Gu. Adaptable platform for interactive swarm robotics (APIS): A human-swarm interaction research testbed. In *Proceedings of the 19th International Conference on Advanced Robotics*, pages 720–726, 2019.

[83] Elisa Donati, Godfried J van Vuuren, Katsuaki Tanaka, Donato Romano, Thomas Schmickl, and Cesare Stefanini. amussels: Diving and anchoring in a new bio-inspired under-actuated robot class for long-term environmental exploration and monitoring. In *Proceedings of the Conference Towards Autonomous Robotic Systems*, pages 300–314. Springer, 2017.

[84] Marco Dorigo and Mauro Birattari. Ant colony optimization. In *Encyclopedia of machine learning*, pages 36–39. Springer, 2011.

[85] Marco Dorigo, Mauro Birattari, and Manuele Brambilla. Swarm robotics. *Scholarpedia*, 9(1):1463, 2014.

[86] Marco Dorigo, Mauro Birattari, et al. Swarm intelligence. *Scholarpedia*, 2(9):1462, 2007.

[87] Marco Dorigo, Mauro Birattari, and Thomas Stutzle. Ant colony optimization. *IEEE computational intelligence magazine*, 1(4):28–39, 2006.

[88] Marco Dorigo, Mauro Birattari, and Thomas Stützle. Ant colony optimization: A computational intelligence technique. *IEEE Computational Intelligence Magazine*, 1(4):28–39, 2006.

[89] Marco Dorigo, Dario Floreano, Luca Maria Gambardella, Francesco Mondada, Stefano Nolfi, Tarek Baaboura, Mauro Birattari, Michael Bonani, Manuele Brambilla, Arne Brutschy, et al. Swarmanoid: A Novel Concept for the Study of Heterogeneous Robotic Swarms. *IEEE Robotics & Automation Magazine*, 20(4):60–71, 2013.

[90] Marco Dorigo and Thomas Stützle. *Ant Colony Optimization.* A Bradford Book, The MIT Press, 2004.

[91] Marco Dorigo, Guy Theraulaz, and Vito Trianni. Reflections on the future of swarm robotics. *Science robotics*, 5(49), December 2020.

[92] Barzin Doroodgar, Maurizio Ficocelli, Babak Mobedi, and Goldie Nejat. The search for survivors: Cooperative human-robot interaction in search and rescue environments using semi-autonomous robots. In *Proceedings of the IEEE International Conference on Robotics and Automation*, pages 2858–2863, 2010.

[93] Alexey Dosovitskiy, German Ros, Felipe Codevilla, Antonio Lopez, and Vladlen Koltun. Carla: An open urban driving simulator. In *Conference on robot learning*, pages 1–16. PMLR, 2017.

[94] Miguel Duarte, Vasco Costa, Jorge Gomes, Tiago Rodrigues, Fernando Silva, Sancho Moura Oliveira, and Anders Lyhne Christensen. Evolution of Collective Behaviors for a Real Swarm of Aquatic Surface Robots. *PloS one*, 11(3):25, 2016.

[95] Frederick Ducatelle, Gianni A Di Caro, Alexander Förster, Michael Bonani, Marco Dorigo, Stéphane Magnenat, Francesco Mondada, Rehan O'Grady, Carlo Pinciroli, Philippe Rétornaz, et al. Cooperative navigation in robotic swarms. *Swarm Intelligence*, 8(1):1–33, 2014.

[96] David Duff, Mark Yim, and Kimon Roufas. Evolution of Polybot: A Modular Reconfigurable Robot. In *Proceedings of the Harmonic Drive Intelligent Symposium*, 2001.

[97] Wilfried Elmenreich and Hermann de Meer. Self-organizing networked systems for technical applications: A discussion on open issues. In *Proceedings of the 3rd International Workshop on Self-Organizing Systems*, pages 1–9. Springer, 2008.

[98] Wilfried Elmenreich, Raissa D'Souza, Christian Bettstetter, and Hermann de Meer. A survey of models and design methods for self-organizing networked systems. In *Proceedings of the 4th International Workshop on Self-Organizing Systems*, volume LNCS 5918, page 37–49. Springer, 2009.

[99] Wilfried Elmenreich, Tobias Ibounig, and Istvan Fehervari. Robustness versus Performance in Sorting and Tournament Algorithms. *Acta Polytecnica*, 6(5):7–18, 2009.

[100] Wilfried Elmenreich and Gernot Klingler. Genetic Evolution of a Neural Network for the Autonomous Control of a Four-Wheeled Robot. In *Proceedings of the 6th Mexican International Conference on Artificial Intelligence - Special Session (MICAI)*, pages 396–406, 2007.

[101] Wilfried Elmenreich, Alexander Schnabl, and Melanie Schranz. An artificial hormone-based algorithm for production scheduling from the bottom-up. In *Proceedings of the 13th International Conference on Agents and Artificial Intelligence*. SciTePress, 2021.

[102] Wilfried Elmenreich and Stefan Schuster. Demand response by decentralized device control based on voltage level. In *Proceedings of the 7th International Workshop on Self-Organizing Systems* pages 186–189. Springer Verlag, 2013.

[103] Thomas Eltz. Tracing pollinator footprints on natural flowers. *Journal of Chemical Ecology*, 32:907–915, 2006.

[104] Sebastian Engell, Radoslav Paulen, Michel A Reniers, Christian Sonntag, and Haydn Thompson. Core research and innovation areas in cyber-physical systems of systems. In *Proceedings of the International Workshop on Design, Modeling, and Evaluation of Cyber Physical Systems*, pages 40–55. Springer, 2015.

[105] Adam Erskine, Thomas Joyce, and J. Michael Herrmann. Stochastic stability of particle swarm optimisation. *Swarm Intelligence*, 11(3):295–315, 2017.

[106] Kasra Eshaghi, Yuchen Li, Zendai Kashino, Goldie Nejat, and Beno Benhabib. mroberto 2.0 – an autonomous millirobot with enhanced locomotion for swarm robotics. *IEEE Robotics and Automation Letters*, 5(2):962–969, 2020.

[107] Hadi Eskandar, Ali Sadollah, Ardeshir Bahreininejad, and Mohd Hamdi. Water cycle algorithm – a novel metaheuristic optimization method for solving constrained engineering optimization problems. *Computers & Structures*, 110:151–166, 2012.

[108] Tingxiang Fan, Pinxin Long, Wenxi Liu, and Jia Pan. Distributed multi-robot collision avoidance via deep reinforcement learning for navigation in complex scenarios. *The International Journal of Robotics Research*, 39(7):856–892, 2020.

[109] Jolyon J Faria, John RG Dyer, Colin R Tosh, and Jens Krause. Leadership and social information use in human crowds. *Animal Behaviour*, 79(4):895 – 901, 2010.

[110] B. G. Farley and W. Clark. Simulation of self-organizing systems by digital computer. *Transactions of the IRE Professional Group on Information Theory*, 4(4):76–84, 1954.

[111] István Fehérvári and Wilfried Elmenreich. Evolving neural network controllers for a team of self-organizing robots. *Journal of Robotics*, 2010.

[112] István Fehérvári and Wilfried Elmenreich. Evolution as a tool to design self-organizing systems. In *Proceedings of the IFIP 7th International Workshop on Self-Organizing Systems*, pages 139–144, 2013.

[113] Ouajdi Felfoul, Mahmood Mohammadi, Samira Taherkhani, Dominic De Lanauze, Yong Zhong Xu, Dumitru Loghin, Sherief Essa, Sylwia Jancik, Daniel Houle, Michel Lafleur, et al. Magneto-aerotactic bacteria deliver drug-containing nanoliposomes to tumour hypoxic regions. *Nature Nanotechnology*, 11(11), 7 p., 2016.

[114] Jean-Claude Fernandez, Laurent Mounier, and Cyril Pachon. A model-based approach for robustness testing. In *IFIP International Conference on Testing of Communicating Systems*, pages 333–348. Springer, 2005.

[115] Eliseo Ferrante, Edgar Duéñez Guzmán, Ali Emre Turgut, and Tom Wenseleers. Geswarm: Grammatical evolution for the automatic synthesis of collective behaviors in swarm robotics. In *Proceedings of the 15th Annual Conference on Genetic and Evolutionary Computation*, pages 17–24, 2013.

[116] Eliseo Ferrante, Ali Emre Turgut, Marco Dorigo, and Cristian Huepe. Collective motion dynamics of active solids and active crystals. *New Journal of Physics*, 15(9):095011, 2013.

[117] Eliseo Ferrante, Ali Emre Turgut, Marco Dorigo, and Cristián Huepe. Elasticity-driven collective motion in active solids and active crystals. *arXiv preprint arXiv:1301.2620*, 2013.

[118] Eliseo Ferrante, Ali Emre Turgut, Cristián Huepe, Alessandro Stranieri, Carlo Pinciroli, and Marco Dorigo. Self-organized flocking with a mobile robot swarm: a novel motion control method. *Adaptive Behavior*, 20(6):460–477, 2012.

[119] Eliseo Ferrante, Ali Emre Turgut, Alessandro Stranieri, Carlo Pinciroli, Mauro Birattari, and Marco Dorigo. A self-adaptive communication strategy for flocking in stationary and non-stationary environments: complete data. *Supplementary information http://iridia. ulb. ac. be/supp/IridiaSupp2011-025/. Accessed*, 6, 2013.

[120] Stanley Fields. Phermone response in yeast. *Trends in Biochemical Sciences*, 15(7):270–273, 1990.

[121] Apache Flink. `https://flink.apache.org/`. [Online; accessed 8-August-2024].

[122] Dario Floreano and Claudio Mattiussi. *Bio-Inspired Artificial Intelligence: Theories, Methods, and Technologies.* MIT Press, 2008.

[123] National Science Foundation. Cyber Physical Systems. `https://www.nsf.gov/funding/pgm_summ.jsp?pims_id=503286`. [Online; accessed 15-March-2018].

[124] Gianpiero Francesca, Manuele Brambilla, Arne Brutschy, Vito Trianni, and Mauro Birattari. Automode: A novel approach to the automatic design of control software for robot swarms. *Swarm Intelligence*, 8(2):89–112, 2014.

[125] Ryusuke Fujisawa, Shigeto Dobata, Ken Sugawara, and Fumitoshi Matsuno. Designing pheromone communication in swarm robotics: Group foraging behavior mediated by chemical substance. *Swarm Intelligence*, 8(3):227–246, 2014.

[126] David Furlonger. Swarm intelligence: From smart cars to smart traffic, BusinessLIVE, August 2017. https://www.businesslive.co.za/bd/life/motoring/2017-08-14-swarm-intelligence-from-smart-cars-to-smart-traffic, 2017. [Online; accessed 27-September-2019].

[127] Epic Games. Unity Technologies. Unity. Available at: https://unity.com/de [Accessed: 26 March 2023].

[128] Epic Games. Unreal Engine. https://www.unrealengine.com/, 2024. [Online; accessed 23-March-2023].

[129] X. Z. Gao, V. Govindasamy, H. Xu, K. Wang, and K. Zenger. Harmony search method: Theory and applications. *Computational Intelligence and Neuroscience*, 258491, April 2015.

[130] Michael R Garey, David S Johnson, and Ravi Sethi. The complexity of flowshop and jobshop scheduling. *Mathematics of Operations Research*, 1(2):117–129, 1976.

[131] Simon Garnier, Jacques Gautrais, and Guy Theraulaz. The biological principles of swarm intelligence. *Swarm Intelligence*, 1(1):3–31, 2007.

[132] Simon Garnier, Christian Jost, Jacques Gautrais, Masoud Asadpour, Gilles Caprari, Raphaël Jeanson, Anne Grimal, and Guy Theraulaz. The Embodiment of Cockroach Aggregation Behavior in a Group of Micro-robots. *Artificial life*, 14(4):387–408, 2008.

[133] Simon Garnier, Faben Tache, Maud Combe, Anne Grimal, and Guy Theraulaz. Alice in pheromone land: An experimental setup for the study of ant-like robots. In *2007 IEEE Swarm Intelligence Symposium*, pages 37–44. IEEE, 2007.

[134] Andrea Gasparri, Giuseppe Oriolo, Attilio Priolo, and Giovanni Ulivi. A swarm aggregation algorithm for multi-robot systems based on local interaction. In *Proceedings of the IEEE International Conference on Control Applications*, pages 1497–1502, 2012.

[135] Alexander J. Gates and Luis M. Rocha. Control of complex networks requires both structure and dynamics. *Nature, Scientific Reports*, 6(24456), 2016.

[136] Veysel Gazi and Kevin M. Passino. Stability analysis of social foraging swarms. *IEEE Transactions on Systems, Man, and Cybernetics: Part B*, 34(1):539–557, 2004.

[137] GCtronic Sagl. Elisa-3 robots - 38 units with obstacle avoidance. https://www.youtube.com/watch?v=WDxfIFhpm1g, 2014. [Online; accessed 30-January-2020].

[138] Hwaiu Geng, editor. *Semiconductor Manufacturing Handbook*. McGraw-Hill Education, 2018.

[139] Edd Gent. Ever dream of controlling robot swarms? This new virtual reality headset could help. *Science*, 2019. `https://www.science.org/content/article/ever-dream-controlling-robot-swarms-new-virtual-reality-headset-could-help`,[Online; accessed 21-December-2023].

[140] Carlos Gershenson. *Design and control of self-organizing systems*. CopIt Arxives, 2007.

[141] Mohammadreza Ghorbaniparvar. Survey on forced oscillations in power system. *Journal of Modern Power Systems and Clean Energy*, 5:671–682, June 2017.

[142] S. N. Givigi and H. M. Schwartz. A game theoretic approach to swarm robotics. *Applied Bionics and Biomechanics*, 3(3):131–142, 2006.

[143] Alan Godoy, Pedro Tabacof, and Fernando J. Von Zuben. The role of the interaction network in the emergence of diversity of behavior. *PLoS ONE*, 12(2):e0172073, 2017.

[144] Jorge Gomes, Paulo Urbano, and Anders Lyhne Christensen. Evolution of swarm robotics systems with novelty search. *Swarm Intelligence*, 7(2):115–144, 2013.

[145] David Green, Aldeida Aleti, and Julian Garcia. The nature of nature: Why nature-inspired algorithms work. In Srikanta Patnaik, Xin-She Yang, and Kazumi Nakamatsu, editors, *Nature-Inspired Computing and Optimization: Theory and Applications*, pages 1–27. Springer, 2017.

[146] Calvin Gregory and Andrew Vardy. microUSV: A low-cost platform for indoor marine swarm robotics research. *HardwareX*, 7:e00105, 2020.

[147] Kay S. Grennan, Chao Chen, Elliot S. Gershon, and Chunyu Liu. Molecular network analysis enhances understanding of the biology of mental disorders. *Bioessays*, 36(6):606–616, 2014.

[148] Boris Gromov, Luca Gambardella, and Gianni A. Di Caro. Wearable multi-modal interface for human multi-robot interaction. In *Proceedings of the 14th IEEE International Symposium on Safety, Security, and Rescue Robotics*, pages 240–245, 2016.

[149] Roderich Groß, Michael Bonani, Francesco Mondada, and Marco Dorigo. Autonomous Self-Assembly in Swarm-Bots. *IEEE Transactions on Robotics*, 22(6):1115–1130, 2006.

[150] D. Grossman, I. S. Aranson, and E. Ben Jacob. Emergence of agent swarm migration and vortex formation through inelastic collisions. *New Journal of Physics*, 10(2):023036, 2008.

[151] Chathika Gunaratne and Ivan Garibay. NL4Py: Agent-based modeling in Python with parallelizable NetLogo workspaces. *SoftwareX*, 16:100801, 2021.

[152] Kexin Guo, Zhirong Qiu, Cunxiao Miao, Abdul Hanif Zaini, Chun-Lin Chen, Wei Meng, and Lihua Xie. Ultra-wideband-based localization for quadcopter navigation. *Unmanned Systems*, 4(01):23–34, 2016.

[153] Heiko Hamann. *Swarm robotics: A formal approach.* Springer, 2018.

[154] Kyle Robert Harrison, Andries P. Engelbrecht, and Beatrice M. Ombuki-Berman. Self-adaptive particle swarm optimization: A review and analysis of convergence. *Swarm Intelligence*, 12(3):187–226, 2018.

[155] Aboul Ella Hassanien and Eid Alamry. *Swarm Intelligence: Principles, Advances, and Applications.* CRC Press, 2015.

[156] Sabine Hauert, Jean-Christophe Zufferey, and Dario Floreano. Evolved swarming without positioning information: An application in aerial communication relay. *Autonomous Robots*, 26(1):21–32, 2009.

[157] Apache HBase. `https://hbase.apache.org/`. [Online; accessed 8-August-2024].

[158] Anni Heckert. *Entwicklung eines dynamischen Modells und Parameterschätzung für den FINken 3 Quadkopter.* Bachelor's thesis, Fakultät für Elektrotechnik und Informationstechnik, Otto-von-Guericke-Universitat Magdeburg, 2016.

[159] Herianto and D. Kurabayashi. Realization of an artificial pheromone system in random data carriers using rfid tags for autonomous navigation. In *2009 IEEE International Conference on Robotics and Automation*, pages 2288–2293, May 2009.

[160] Francis Heylighen. The science of self-organization and adaptivity. *The Encyclopedia of Life Support Systems*, 5(3):253–280, 2001.

[161] Francis Heylighen. Complexity and self-organization, 2008.

[162] Nicholas R. Hoff, Amelia Sagoff, Robert J. Wood, and Radhika Nagpal. Two foraging algorithms for robot swarms using only local communication. In *Proceedings of the IEEE International Conference on Robotics and Biomimetics*, pages 123–130, 2010.

[163] John Henry Holland et al. *Adaptation in natural and artificial systems: an introductory analysis with applications to biology, control, and artificial intelligence.* MIT press, 1992.

[164] Richard Holzer and Hermann De Meer. On modeling of self-organizing systems. In *Proceedings of the 2nd International Conference on Autonomic Computing and Communication Systems*, pages 1–6, 2008.

[165] Richard Holzer and Hermann De Meer. Quantitative modeling of self-organizing properties. In *International Workshop on Self-Organizing Systems*, pages 149–161. Springer, 2009.

[166] Matthew B Hoy. Alexa, siri, cortana, and more: An introduction to voice assistants. *Medical Reference Services Quarterly*, 37(1):81–88, 2018.

[167] Allen Hsu, Huihua Zhao, Martin Gaudreault, Annjoe Wong Foy, and Ron Pelrine. Magnetic milli-robot swarm platform: A safety barrier certificate enabled, low-cost test bed. *IEEE Robotics and Automation Letters*, 5(2):2913–2920, 2020.

[168] Junyan Hu, Parijat Bhowmick, Farshad Arvin, Alexander Lanzon, and Barry Lennox. Cooperative control of heterogeneous connected vehicle platoons: An adaptive leader-following approach. *IEEE Robotics and Automation Letters*, 5(2):977–984, 2020.

[169] Junyan Hu, Parijat Bhowmick, Inmo Jang, Farshad Arvin, and Alexander Lanzon. A decentralized cluster formation containment framework for multirobot systems. *IEEE Transactions on Robotics*, 2021.

[170] Junyan Hu, Hanlin Niu, Joaquin Carrasco, Barry Lennox, and Farshad Arvin. Voronoi-based multi-robot autonomous exploration in unknown environments via deep reinforcement learning. *IEEE Transactions on Vehicular Technology*, 69(12):14413–14423, 2020.

[171] Junyan Hu, Ali Emre Turgut, Tomáš Krajník, Barry Lennox, and Farshad Arvin. Occlusion-based coordination protocol design for autonomous robotic shepherding tasks. *IEEE Transactions on Cognitive and Developmental Systems*, 2020.

[172] Junyan Hu, Ali Emre Turgut, Barry Lennox, and Farshad Arvin. Robust formation coordination of robot swarms with nonlinear dynamics and unknown disturbances: Design and experiments. *IEEE Transactions on Circuits and Systems II: Express Briefs*, 2021.

[173] Ye Huang, Amos Brocco, Pierre Kuonen, Michèle Courant, and Béat Hirsbrunner. Smartgrid: A fully decentralized grid scheduling framework supported by swarm intelligence. In *Proceedings of the 7th International Conference on Grid and Cooperative Computing*, pages 160–168, 2008.

[174] Mimmo Iannelli and Andrea Pugliese. *An Introduction to Mathematical Population Dynamics: A long the trail of Volterra and Lotka*. Springer, 2014.

[175] Infineon Technologies AG. `https://www.infineon.com/` [Online; accessed 22-March-2024].

[176] Malek Itani, Tuochao Chen, Takuya Yoshioka, and Shyamnath Gollakota. Creating speech zones with self-distributing acoustic swarms. *Nature Communications*, 14(1):5684, 2023.

[177] Duncan E. Jackson and Francis L.W. Ratnieks. Communication in ants. *Current Biology*, 16(15):R570 – R574, 2006.

[178] Inmo Jang, Junyan Hu, Farshad Arvin, Joaquin Carrasco, and Barry Lennox. Omnipotent virtual giant for remote human–swarm interaction. In *2021 30th IEEE International Conference on Robot & Human Interactive Communication (RO-MAN)*, pages 488–494. IEEE, 2021.

[179] Fredrik Jansson, Matthew Hartley, Martin Hinsch, Ivica Slavkov, Noemí Carranza, Tjelvar SG Olsson, Roland M Dries, Johanna H Grönqvist, Athanasius FM Marée, James Sharpe, et al. Kilombo: a kilobot simulator to enable effective research in swarm robotics. *arXiv preprint arXiv:1511.04285*, 2015.

[180] Midhat Jdeed, Arthur Pitman, and Wilfried Elmenreich. Spiderino - a low cost robot for swarm research and educational purposes, technical documentation. Technical report, Institute of Networked and Embedded Systems, Alpen-Adria-Universität Klagenfurt, Austria, 2018.

[181] Midhat Jdeed, Sergii Zhevzhyk, Florian Steinkellner, and Wilfried Elmenreich. Spiderino: A low-cost Robot for Swarm Research and Educational Purposes. In *Proceedings of the 13th Workshop on Intelligent Solutions in Embedded Systems*, pages 35–39, 2017.

[182] Shu Jiang and Ronald C Arkin. Mixed-initiative human-robot interaction: Definition, taxonomy, and survey. In *Proceedings of the IEEE International Conference on Systems, Man, and Cybernetics*, pages 954–961, 2015.

[183] Yaochu Jin and Jurgen Branke. Evolutionary optimization in uncertain environments-a survey. *IEEE Transactions on Evolutionary Computation*, 9(3):303–317, 2005.

[184] Martin Jinek, Krzysztof Chylinski, Ines Fonfara, Michael Hauer, Jennifer A. Doudna, and Emmanuelle Charpentier. A programmable dual-RNA-guided DNA endonuclease in adaptive bacterial immunity. *Science*, 337(6096):816–821, 2012.

[185] Geraint Jones, Nadia Berthouze, Roman Bielski, and Simon Julier. Towards a situated, multimodal interface for multiple UAV control. In *Proceedings of the IEEE International Conference on Robotics and Automation*, pages 1739–1744, 2010.

[186] Simon Jones, Emma Milner, Mahesh Sooriyabandara, and Sabine Hauert. Dots: An open testbed for industrial swarm robotic solutions. *arXiv preprint arXiv:2203.13809*, 2022.

[187] Simon Jones, Matthew Studley, Sabine Hauert, and Alan Frank Thomas Winfield. A two Teraflop Swarm. *Frontiers in Robotics and AI*, 5:11, 2018.

[188] Visakan Kadirkamanathan, Kirusnapillai Selvarajah, and Peter J. Fleming. Stability analysis of the particle dynamics in particle swarm optimizer. *IEEE Transactions on Evolutionary Computation*, 10(3), 2006.

[189] Apache Kafka. `https://kafka.apache.org`. [Online; accessed 8-August-2024].

[190] Fumio Kanehiro, Hirohisa Hirukawa, and Shuuji Kajita. Openhrp: Open architecture humanoid robotics platform. *The International Journal of Robotics Research*, 23(2):155–165, 2004.

[191] Minoru Kanehisa, Susumu Goto, Miho Furumichi, Mao Tanabe, and Mika Hirakawa. KEGG for representation and analysis of molecular networks involving diseases and drugs. *Nucleic Acids Research*, 38(suppl_1):D355–D360, 2010.

[192] Chang-Kwon Kang. Marsbee - Swarm of Flapping Wing Flyers for Enhanced Mars Exploration. `https://www.nasa.gov/directorates/spacetech/niac/2018_Phase_I_Phase_II/Marsbee_Swarm_of_Flapping_Wing_Flyers_for_Enhanced_Mars_Exploration`, 2018. [Online; accessed 05-September-2019].

[193] Akbar Karimi, Hadi Nobahari, and Patrick Siarry. Continuous ant colony system and tabu search algorithms hybridized for global minimization of continuous multi-minima functions. *Computational Optimization and Applications*, 45(3):639–661, 2010.

[194] P. Karlson and M Lüscher. 'Pheromones': A New Term for a Class of Biologically Active Substances. *Nature*, 183:55–56, 1959.

[195] Michael N. Katehakis and Arthur F. Veinott. The multi-armed bandit problem: Decomposition and computation. *Mathematics of Operations Research*, 12(2):262–268, 1987.

[196] Ali Kaveh and Neda Farhoudi. A new optimization method: Dolphin echolocation. *Advances in Engineering Software*, 59:53–70, 2013.

[197] James Kennedy and Russell C. Eberhart. Particle swarm optimization. In *Proceedings of the IEEE International Conference on Neural Networks*, pages 1942–1948, 1995.

[198] Serge Kernbach, Dagmar Häbe, Olga Kernbach, Ronald Thenius, Gerald Radspieler, Toshifumi Kimura, and Thomas Schmickl. Adaptive Collective Decision-making in Limited Robot Swarms without Communication. *The International Journal of Robotics Research*, 32(1):35–55, 2013.

[199] Serge Kernbach, Eugen Meister, Florian Schlachter, Kristof Jebens, Marc Szymanski, Jens Liedke, Davide Laneri, Lutz Winkler, Thomas Schmickl, Ronald Thenius, et al. Symbiotic Robot Organisms: REPLICATOR and SYMBRION Projects. In *Proceedings of the 8th Workshop on Performance Metrics for Intelligent Systems*, pages 62–69, 2008.

[200] Alexander Kettler. Cleaning up with a Swarm of Robots. `https://www.youtube.com/watch?v=FqMP_AIkDj8`, 2012. [Online; accessed 05-February-2020].

[201] Alexander Kettler, Marc Szymanski, and Heinz Wörn. The Wanda Robot and its Development System for Swarm Algorithms. In *Advances in Autonomous Mini Robots*, pages 133–146. Springer, 2012.

[202] Ali Abdul Khaliq and Alessandro Saffiotti. Stigmergy at work: Planning and navigation for a service robot on an rfid floor. In *2015 IEEE International Conference on Robotics and Automation (ICRA)*, pages 1085–1092, May 2015.

[203] Elnaz Khatmi, Wilfried Elmenreich, Kristina Wogatai, Melanie Schranz, Martina Umlauft, Walter Laure, and Andreas Wuttei. Swarm intelligence layer to control autonomous agents (SWILT). In *Proceedings of the Research Project Showcase at Software Technologies: Applications and Foundations (STAF-RPS19)*, 2019.

[204] Kicksat. Kicksat: A tiny open source spacecraft project. `https://kicksat.github.io/`. [Online; accessed 9-July-2019].

[205] Hannes Kirchhoff, Noara Kebir, Kirsten Neumann, Peter W. Heller, and Kai Strunz. Developing mutual success factors and their application to swarm electrification: microgrids with 100 % renewable energies in the Global South and Germany. *Journal of Cleaner Production*, 128:190–200, 2016.

[206] Scott Kirkpatrick, C Daniel Gelatt, and Mario P Vecchi. Optimization by simulated annealing. In *Readings in Computer Vision*, pages 606–615. Elsevier, 1987.

[207] Zalan Kiszli, Seongin Na, and Farshad Arvin. Toward a Myriad Robot Swarm Aggregation. In *7th International Conference on Control and Robotics Engineering*. IEEE, 2022.

[208] John Klingner, Anshul Kanakia, Nicholas Farrow, Dustin Reishus, and Nikolaus Correll. A Stick-slip Omnidirectional Powertrain for low-cost Swarm Robotics: Mechanism, Calibration, and Control. *IEEE International Conference on Intelligent Robots and Systems*, pages 846–851, 2014.

[209] Nathan Koenig and Andrew Howard. Design and use paradigms for gazebo, an open-source multi-robot simulator. In *2004 IEEE/RSJ International Conference on Intelligent Robots and Systems (IROS)(IEEE Cat. No. 04CH37566)*, volume 3, pages 2149–2154. IEEE, 2004.

[210] Andreas Kolling, Phillip Walker, Nilanian Chakraborty, Katia Sycara, and Michael Lewis. Human interaction with robot swarms: A survey. *IEEE Transactions on Human-Machine Systems*, 46(1):9–26, 2016.

[211] Shigeru Kondo and Takashi Miura. Reaction-diffusion model as a framework for understanding biological pattern formation. *science*, 329(5999):1616–1620, 2010.

[212] Tomáš Krajník, Matías Nitsche, Jan Faigl, Petr Vaněk, Martin Saska, Libor Přeučil, Tom Duckett, and Marta Mejail. A practical multirobot localization system. *Journal of Intelligent & Robotic Systems*, 76(3-4):539–562, 2014.

[213] Jens Krause, Graeme D Ruxton, and Stefan Krause. Swarm intelligence in animals and humans. *Trends in Ecology & Evolution*, 25(1):28 – 34, 2010.

[214] Jonas Krause, Jelson Cordeiro, Rafael Stubs Parpinelli, and Heitor Silverio Lopes. A Survey of Swarm Algorithms Applied to Discrete Optimization. In Xin-She Yang, Zhihua Cui, Renbin Xiao, Amir Hossein Gandomi, and Mehmet Karamanoglu, editors, *Swarm Intelligence and Bio-Inspired Computation*, pages 169–191. Elsevier, 2013.

[215] Stefan Krause, Richard James, Jolyon J Faria, Graeme D Ruxton, and Jens Krause. Swarm intelligence in humans: Diversity can trump ability. *Animal Behaviour*, 81(5):941 – 948, 2011.

[216] Geert-Jan M Kruijff, M Janíček, Shanker Keshavdas, Benoit Larochelle, Hendrik Zender, Nanja JJM Smets, Tina Mioch, Mark A Neerincx, Jurriaan Van Diggelen, Francis Colas, et al. Experience in system design for human-robot teaming in urban search and rescue. In *Field and Service Robotics*, pages 111–125, 2014.

[217] Haruhisa Kurokawa, Akiya Kamimura, Eiichi Yoshida, Kohji Tomita, Shigeru Kokaji, and Satoshi Murata. M-TRAN II: Metamorphosis from a Four-legged Walker to a Caterpillar. In *Proceedings of the IEEE/RSJ International Conference on Intelligent Robots and Systems*, volume 3, pages 2454–2459, 2003.

[218] Haruhisa Kurokawa, Kohji Tomita, Akiya Kamimura, Shigeru Kokaji, Takashi Hasuo, and Satoshi Murata. Distributed Self-reconfiguration of M-TRAN III Modular Robotic System. *The International Journal of Robotics Research*, 27(3-4):373–386, 2008.

[219] Chang kwon Kang. Marsbee: Swarm of flapping wing flyers for enhanced Mars exploration. `https://www.nasa.gov/directorates/spacetech/niac/2018_Phase_I_Phase_II/Marsbee_Swarm_of_Flapping_Wing_Flyers_for_Enhanced_Mars_Exploration`, 2018. [Online; accessed 9-July-2019].

[220] George Kyriakarakos and George Papadakis. Multispecies Swarm Electrification for Rural Areas of the Developing World. *Applied Sciences*, 9(19):3992, 2019.

[221] Jean-Claude Laprie. From dependability to resilience. In *38th IEEE/IFIP Int. Conf. On dependable systems and networks*, pages G8–G9, 2008.

[222] Eugene L Lawler, Jan Karel Lenstra, Alexander HG Rinnooy Kan, and David B Shmoys. Sequencing and scheduling: Algorithms and complexity. *Handbooks in Operations Research and Management Science*, 4:445–522, 1993.

[223] Mathieu Le Goc, Lawrence H Kim, Ali Parsaei, Jean-Daniel Fekete, Pierre Dragicevic, and Sean Follmer. Zooids: Building Blocks for Swarm User Interfaces. In *Proceedings of the 29th Annual Symposium on User Interface Software and Technology*, pages 97–109, 2016.

[224] Edward A Lee. Cyber physical systems: Design challenges. In *Proceedings of the 11th IEEE International Symposium on Object Oriented Real-Time Distributed Computing*, pages 363–369, 2008.

[225] Edward A. Lee. The past, present and future of cyber-physical systems: A focus on models. *Transactions on Cyber-Physical Systems*, 1(1):3:1–3:26, February 2017.

[226] Edward Ashford Lee and Sanjit A Seshia. *Introduction to embedded systems: A cyber-physical systems approach*. MIT Press, 2016.

[227] Lin Li, Ruochen Hao, Wanjing Ma, Xinzhou Qi, and Chenxue Diao. Swarm intelligence based algorithm for management of autonomous vehicles on arterials. Technical report, SAE, 2018.

[228] Shimin Li, Huiling Chen, Mingjing Wang, Ali Asghar Heidari, and Seyedali Mirjalili. Slime mould algorithm: A new method for stochastic optimization. *Future Generation Computer Systems*, 111:300–323, 2020.

[229] Sean Lim, Shiyi Wang, Barry Lennox, and Farshad Arvin. Beeground-an open-source simulation platform for large-scale swarm robotics applications. In *2021 7th International Conference on Automation, Robotics and Applications (ICARA)*, pages 75–79. IEEE, 2021.

[230] Marcelo A. Limeira, Luis Piardi, Vivian Cremer Kalempa, André Schneider de Oliveira, and Paulo Leitão. Wsbot: A tiny, low-cost swarm robot for experimentation on industry 4.0. In *2019 Latin American Robotics Symposium (LARS), 2019 Brazilian Symposium on Robotics (SBR) and 2019 Workshop on Robotics in Education (WRE)*, pages 293–298, 2019.

[231] Tian Liu, Xuelong Sun, Cheng Hu, Qinbing Fu, and Shigang Yue. A multiple pheromone communication system for swarm intelligence. *IEEE Access*, 9:148721–148737, 2021.

[232] Yang-Yu Liu, Jean-Jacques Slotine, and Albert-László Barabási. Observability of complex systems. *Proceedings of the National Academy of Sciences*, 110(7):2460–2465, 2013.

[233] Zheyu Liu, Craig West, Barry Lennox, and Farshad Arvin. Local bearing estimation for a swarm of low-cost miniature robots. *Sensors*, 20(11):3308, 2020.

[234] Yuri K. Lopes, Stefan M. Trenkwalder, André B. Leal, Tony J. Dodd, and Roderich Groß. Supervisory control theory applied to swarm robotics. *Swarm Intelligence*, 10(1):65–97, 2016.

[235] Manuel López-Ibáñez and Thomas Stützle. An experimental analysis of design choices of multi-objective ant colony optimization algorithms. *Swarm Intelligence*, 6(3):207–232, 2012.

[236] Sean Luke, Claudio Cioffi-Revilla, Liviu Panait, and Keith Sullivan. MASON: A new multi-agent simulation toolkit. In *Proceedings of the 2004 SwarmFest Workshop*, pages 316–327, Ann Arbor, Michigan, USA, May 2004.

[237] Arun Mahadev, Dominik Krupke, Sandor P. Fekete, and Aaron T. Becker. Mapping and coverage with a particle swarm controlled by uniform inputs. In *Proceedings of the IEEE/RSJ International Conference on Intelligent Robots and Systems*, pages 1097–1104, 2017.

[238] Lorenzo Marconi, Claudio Melchiorri, Michael Beetz, Dejan Pangercic, Roland Siegwart, Stefan Leutenegger, Raffaella Carloni, Stefano Stramigioli, Herman Bruyninckx, Patrick Doherty, et al. The SHERPA project: Smart collaboration between humans and ground-aerial robots for improving rescuing activities in alpine environments. In *Proceedings of the IEEE International Symposium on Safety, Security, and Rescue Robotics*, pages 1–4, 2012.

[239] David Masad and Jacqueline Kazil. Mesa: an agent-based modeling framework. In *14th PYTHON in Science Conference*, volume 2015, pages 53–60. Citeseer, 2015.

[240] Ralf Mayet, Jonathan Roberz, Thomas Schmickl, and Karl Crailsheim. Antbots: A feasible visual emulation of pheromone trails for swarm robots. In *Proceedings of the International Conference on Swarm Intelligence*, pages 84–94. Springer, 2010.

[241] Cassandra McCord, Jorge Pena Queralta, Tuan Nguyen Gia, and Tomi Westerlund. Distributed progressive formation control for multi-agent systems: 2d and 3d deployment of uavs in ros/gazebo with rotors. In *2019 European Conference on Mobile Robots (ECMR)*, pages 1–6. IEEE, 2019.

[242] James McLurkin, Andrew J Lynch, Scott Rixner, Thomas W Barr, Alvin Chou, Kathleen Foster, and Siegfried Bilstein. A low-cost Multi-robot System for Research, Teaching, and Outreach. In *Distributed Autonomous Robotic Systems*, pages 597–609. Springer, 2013.

[243] Andrew McNabb, Matthew Gardner, and Kevin Seppi. An exploration of topologies and communication in large particle swarms. In *Proceedings of the IEEE Congress on Evolutionary Computation*, pages 712–719, 2009.

[244] Andreas M Menzel and Takao Ohta. Soft deformable self-propelled particles. *EPL (Europhysics Letters)*, 99(5):58001, 2012.

[245] Vera Mersheeva. *UAV Routing Problem for Area Monitoring in a Disaster Situation.* Ph.d. dissertation, Universit́at Klagenfurt, 2015. Dissertation.

[246] Olivier Michel. Cyberbotics ltd. webots™: professional mobile robot simulation. *International Journal of Advanced Robotic Systems*, 1(1):5, 2004.

[247] Seyedali Mirjalili, Seyed Mohammad Mirjalili, and Andrew Lewis. Grey wolf optimizer. *Advances in engineering software*, 69:46–61, 2014.

[248] Melanie Mitchell. *Complexity: A guided tour.* Oxford University Press, 2009.

[249] Francesco Mondada, Michael Bonani, Xavier Raemy, James Pugh, Christopher Cianci, Adam Klaptocz, Stephane Magnenat, Jean-Christophe Zufferey, Dario Floreano, and Alcherio Martinoli. The e-puck, a robot designed for education in engineering. In *Proceedings of the 9th conference on autonomous robot systems and competitions*, pages 59–65, 2009.

[250] Francesco Mondada, Michael Bonani, Fanny Riedo, Manon Briod, Léa Pereyre, Philippe Rétornaz, and Stéphane Magnenat. Bringing robotics to formal education: The thymio open-source hardware robot. *IEEE Robotics & Automation Magazine*, 24(1):77–85, 2017.

[251] Francesco Mondada, Alcherio Martinoli, Nicolaus Correll, Alexey Gribovskiy, José Ignacio Halloy, Roland Siegwart, and Jean-Louis Deneubourg. A general methodology for the control of mixed natural-artificial societies. *Handbook of Collective Robotics*, pages 399–428, 2011.

[252] Francesco Mondada, Giovanni C Pettinaro, Andre Guignard, Ivo W Kwee, Dario Floreano, Jean-Louis Deneubourg, Stefano Nolfi, Luca Maria Gambardella, and Marco Dorigo. Swarm-bot: A new distributed robotic concept. *Autonomous robots*, 17(2-3):193–221, 2004.

[253] Francesco Mondada, Giovanni Cosimo Pettinaro, Ivo W Kwee, André Guignard, Luca Maria Gambardella, Dario Floreano, Stefano Nolfi, Jean-Louis Deneubourg, and Marco Dorigo. SWARM-BOT: A Swarm of Autonomous Mobile Robots with Self-assembling Capabilities. Technical report, ETH-Zürich, 2002.

[254] MongoDB. `https://www.mongodb.com/`. [Online; accessed 8-August-2024].

[255] Roberto Montemanni, Luca Maria Gambardella, Andrea Emilio Rizzoli, and Alberto V Donati. Ant colony system for a dynamic vehicle routing problem. *Journal of Combinatorial Optimization*, 10(4):327–343, 2005.

[256] Honda Motor. Safe swarm. `https://global.honda/innovation/CES/2019/safe_swarm.html`, 2019. [Online; accessed 01-October-2019].

[257] Jean-Baptiste Mouret and Konstantinos Chatzilygeroudis. 20 years of reality gap: a few thoughts about simulators in evolutionary robotics. In *Proceedings of the Genetic and Evolutionary Computation Conference Companion*, pages 1121–1124. ACM, 2017. DOI: 10.1145/3067695.3082052.

[258] Mehdi Moussaïd, Dirk Helbing, Simon Garnier, Anders Johansson, Maud Combe, and Guy Theraulaz. Experimental study of the behavioural mechanisms underlying self-organization in human crowds. *Proceedings of the Royal Society of London B: Biological Sciences*, 276(1668):2755–2762, 2009.

[259] Ahmad Mozaffari, Alireza Fathi, and Saeed Behzadipour. The great salmon run: A novel bio-inspired algorithm for artificial system design and optimisation. *International Journal of Bio-Inspired Computation*, 4(5):286–301, 2012.

[260] Rolex Muceka, Tonny Kukeera, Yunus Alokore, Kebir Noara, and Sebastian Groh. Integrating a Solar PV System with a Household Based Backup Generator for Hybrid Swarm Electrification: A Case Study of Nigeria. In *Africa-EU Renewable Energy Research and Innovation Symposium 2018 (RERIS 2018)*, pages 43–58. Springer, Cham, 2018.

[261] Satoshi Murata, Eiichi Yoshida, Akiya Kamimura, Haruhisa Kurokawa, Kohji Tomita, and Shigeru Kokaji. M-TRAN: Self-reconfigurable Modular Robotic System. *IEEE/ASME Transactions on Mechatronics*, 7(4):431–441, 2002.

[262] Robin R. Murphy, Satoshi Tadokoro, Daniele Nardi, Adam Jacoff, Paolo Fiorini, Howie Choset, and Aydan M. Erkmen. Search and rescue robotics. In B. Siciliano and O. Khatib, editors, *Springer Handbook of Robotics*, pages 1151–1173. Springer Berlin Heidelberg, Berlin, Heidelberg, 2008.

[263] Smiljana Mutic, Yvonne F. Brünner, Rea Rodriguez-Raecke, Martin Wiesmann, and Jessica Freiherr. Chemosensory danger detection in the human brain: Body odor communicating aggression modulates limbic system activation. *Neuropsychologia*, 99(September 2016):187–198, 2017.

[264] Seongin Na, Hanlin Niu, Barry Lennox, and Farshad Arvin. Universal artificial pheromone framework with deep reinforcement learning for robotic systems. In *6th International Conference on Control and Robotics Engineering (ICCRE)*, pages 28–32. IEEE, 2021.

[265] Seongin Na, Hanlin Niu, Barry Lennox, and Farshad Arvin. Bio-inspired collision avoidance in swarm systems via deep reinforcement learning. *IEEE Transactions on Vehicular Technology*, 2022.

[266] Seongin Na, Yiping Qiu, Ali E Turgut, Jiří Ulrich, Tomáš Krajník, Shigang Yue, Barry Lennox, and Farshad Arvin. Bio-inspired artificial pheromone system for swarm robotics applications. *Adaptive Behavior*, 29(4):395–415, 2021.

[267] Seongin Na, Mohsen Raoufi, Ali Emre Turgut, Tomáš Krajník, and Farshad Arvin. Extended artificial pheromone system for swarm robotic applications. In *Artificial Life Conference Proceedings*, pages 608–615. MIT Press One Rogers Street, Cambridge, MA 02142-1209, USA journals-info ..., 2019.

[268] Seongin Na, Tomáš Rouček, Jiří Ulrich, Jan Pikman, Tom s Krajník, Barry Lennox, and Farshad Arvin. Federated reinforcement learning for collective navigation of robotic swarms. *IEEE Transactions on Cognitive and Developmental Systems*, 2023.

[269] Jawad Nagi. *Symbiotic interaction between humans and robot swarms*. PhD thesis, Department of Informatics, University of Lugano, Switzerland, 2016.

[270] Jawad Nagi, Alessandro Giusti, Luca Gambardella, and Gianni A. Di Caro. Human-swarm interaction using spatial gestures. In *Proceedings of the 27th IEEE/RSJ International Conference on Intelligent Robots and Systems*, pages 3834–3841, 2014.

[271] Changjoo Nam and Dylan A. Shell. Assignment algorithms for modeling resource contention in multirobot task allocation. *IEEE Transactions on Automation Science and Engineering*, 12(3):889–900, 2015.

[272] Yadati Narahari, Dinesh Garg, Ramasuri Narayanam, and Hastagiri Prakash. *Game theoretic problems in network economics and mechanism design solutions*. Springer, 2009.

[273] Mark EJ Newman. The structure and function of complex networks. *SIAM review*, 45(2):167–256, 2003.

[274] Claus Ballegaard Nielsen, Peter Gorm Larsen, John Fitzgerald, Jim Woodcock, and Jan Peleska. Systems of systems engineering: Basic concepts, model-based techniques, and research directions. *ACM Computing Surveys*, 48(2), Sep 2015.

[275] Martin A. Nowak, Corina E. Tarnita, and Edward O. Wilson. The evolution of eusociality. *Nature*, 466(7310):1057–1062, Aug 2010.

[276] Olusegun Olorunda and Andries P. Engelbrecht. Measuring exploration/exploitation in particle swarms using swarm diversity. In *2008 IEEE Congress on Evolutionary Computation (IEEE World Congress on Computational Intelligence)*, pages 1128–1134, 2008.

[277] Nestor I. Ospina, Eduardo Mojica-Nava, Luis G. Jaimes, and Juan M. Calderón. Argrohbots: An affordable and replicable ground homogeneous robot swarm testbed. *IFAC-PapersOnLine*, 54(13):256–261, 2021.

[278] Christoph Osterloh, Thilo Pionteck, and Erik Maehle. MONSUN II: A small and Inexpensive AUV for Underwater Swarms. In *Proceedings of the 7th German Conference on Robotics*, pages 1–6, 2012.

[279] Raymond Oung. The Distributed Flight Array: Summary. `https://www.youtube.com/watch?v=fcradVE9uts`, 2013. [Online; accessed 30-January-2020].

[280] Raymond Oung and Raffaello D'Andrea. The Distributed Flight Array. *Mechatronics*, 21(6):908–917, 2011.

[281] Rafael S. Parpinelli and Heitor S. Lopes. New inspirations in swarm intelligence: A survey. *International Journal of Bio-Inspired Computation*, 3(1):1–16, 2011.

[282] Rafael S Parpinelli and Heitor S Lopes. New inspirations in swarm intelligence: A survey. *International Journal of Bio-Inspired Computation*, 3(1):1–16, 2011.

[283] Jayam Patel and Carlo Pinciroli. Improving human performance using mixed granularity of control in multi-human multi-robot interaction. In *2020 29th IEEE International Conference on Robot and Human Interactive Communication (RO-MAN)*, pages 1135–1142. IEEE, 2020.

[284] Jayam Patel, Yicong Xu, and Carlo Pinciroli. Mixed-granularity human-swarm interaction. In *2019 International Conference on Robotics and Automation (ICRA)*, pages 1059–1065. IEEE, 2019.

[285] David Payton, Mike Daily, Regina Estowski, Mike Howard, and Craig Lee. Pheromone robotics. *Autonomous Robots*, 11(3):319–324, 2001.

[286] Kirstin Hagelskjaer Petersen, Radhika Nagpal, and Justin K Werfel. Termes: An Autonomous Robotic System for Three-dimensional Collective Construction. *Robotics: science and systems VII*, 2011.

[287] Daniel Pickem, Paul Glotfelter, Li Wang, Mark Mote, Aaron Ames, Eric Feron, and Magnus Egerstedt. The robotarium: A remotely accessible swarm robotics research testbed. In *2017 IEEE International Conference on Robotics and Automation (ICRA)*, pages 1699–1706. IEEE, 2017.

[288] Daniel Pickem, Myron Lee, and Magnus Egerstedt. The GRITSBot in its Natural Habitat – a Multi-robot Testbed. In *Proceedings of the IEEE International Conference on Robotics and Automation*, pages 4062–4067, 2015.

[289] Carlo Pinciroli and Giovanni Beltrame. Swarm-oriented Programming of Distributed Robot Networks. *Computer*, 49(12):32–41, 2016.

[290] Carlo Pinciroli, Vito Trianni, Rehan O'Grady, Giovanni Pini, Arne Brutschy, Manuele Brambilla, Nithin Mathews, Eliseo Ferrante, Gianni Di Caro, Frederick Ducatelle, et al. Argos: a modular, parallel, multi-engine simulator for multi-robot systems. *Swarm intelligence*, 6(4):271–295, 2012.

[291] G. Porter. Models of Proba-3 designs. `http://www.esa.int/spaceinimages/Images/2016/05/Models_of_Proba-3_designs`, 2016. [Online; accessed 9-July-2019].

[292] Christian Prehofer and Christian Bettstetter. Self-organization in communication networks: principles and design paradigms. *IEEE Communications Magazine*, 43(7):78–85, 2005.

[293] James A. Preiss, Wolfgang Hönig, Gaurav S. Sukhatme, and Nora Ayanian. Crazyswarm: A large Nano-quadcopter Swarm. In *Proceedings of the International Conference on Robotics and Automation*, pages 3299–3304, 2017. Software available at `https://github.com/USC-ACTLab/crazyswarm`.

[294] George Preti, Charles J. Wysocki, Kurt T. Barnhart, Steven J. Sondheimer, and James J. Leyden. Male Axillary Extracts Contain Pheromones that Affect Pulsatile Secretion of Luteinizing Hormone and Mood in Women Recipients1. *Biology of Reproduction*, 68(6):2107–2113, 06 2003.

[295] CPSwarm H2020 project. CPSwarm Modeler. `https://github.com/cpswarm/modelio-cpswarm-modeler`, 2019. [Online; accessed 23-March-2023].

[296] CPSwarm H2020 project. CPSwarm Code Generator. `https://github.com/cpswarm/code-generator`, 2020. [Online; accessed 23-March-2023].

[297] CPSwarm H2020 project. CPSwarm Communication Library. `https://github.com/cpswarm/swarmio`, 2022. [Online; accessed 23-March-2023].

[298] Przemyslaw Prusinkiewicz, Mitra Shirmohammadi, and Faramarz Samavati. L-systems in geometric modeling. *International Journal of Foundations of Computer Science*, 23(01):133–146, 2012.

[299] Amanda Purington, Jessie G Taft, Shruti Sannon, Natalya N Bazarova, and Samuel Hardman Taylor. Alexa is my new bff: Social roles, user satisfaction, and personification of the amazon echo. In *Proceedings of the 2017 CHI Conference Extended Abstracts on Human Factors in Computing Systems*, pages 2853–2859, 2017.

[300] Morgan Quigley, Brian Gerkey, Ken Conley, Josh Faust, Tully Foote, Jeremy Leibs, Eric Berger, Rob Wheeler, and Andrew Ng. ROS: an open-source robot operating system. In *Proceedings of the ICRA Workshop on Open Source Software in Robotics*, May 2009.

[301] Deepak Rai and Kirti Tyagi. Bio-inspired optimization techniques: a critical comparative study. *ACM SIGSOFT Software Engineering Notes*, 38(4):1–7, 2013.

[302] Steven Railsback, Daniel Ayllón, Uta Berger, Volker Grimm, Steven Lytinen, Colin Sheppard, and Jan Christoph Thiele. Improving execution speed of models implemented in netlogo. *Journal of Artificial Societies and Social Simulation*, 2017.

[303] Steven F Railsback and Volker Grimm. *Agent-based and individual-based modeling: a practical introduction.* Princeton university press, "2nd" edition, 2019.

[304] Bhuvaneswari Ramachandran, Sanjeev K. Srivastava, Chris S. Edrington, and David A. Cartes. An intelligent auction scheme for smart grid market using a hybrid immune algorithm. *IEEE Transactions on Industrial Electronics*, 58(10):4603–4612, 2011.

[305] Mohsen Raoufi, Ali Emre Turgut, and Farshad Arvin. Self-organized collective motion with a simulated real robot swarm. In *Proceedings of the Annual Conference Towards Autonomous Robotic Systems*, pages 263–274. Springer, 2019.

[306] Martin Reed, Aliki Yiannakou, and Roxanne Evering. An ant colony algorithm for the multi-compartment vehicle routing problem. *Applied Soft Computing*, 15:169–176, 2014.

[307] Andreagiovanni Reina, Alexander Cope, Eleftherios Nikolaidis, James Marshall, and Chelsea Sabo. Ark: Augmented reality for kilobots. *IEEE Robotics and Automation Letters*, PP:1–1, 05 2017.

[308] Fatemeh Rekabi, Farzad A Shirazi, Mohammad Jafar Sadigh, and Mahmood Saadat. Distributed output feedback nonlinear h∞ formation control algorithm for heterogeneous aerial robotic teams. *Robotics and Autonomous Systems*, 136:103689, 2021.

[309] Fatemeh Rekabi-Bana, Junyan Hu, Tomáš Krajník, and Farshad Arvin. Unified robust path planning and optimal trajectory generation for efficient 3d area coverage of quadrotor uavs. *IEEE Transactions on Intelligent Transportation Systems*, 2023.

[310] Intel Corporation. Intel drone light show breaks guinness world records title at olympic winter games pyeongchang. `https://newsroom.intel.com/news-releases/intel-drone-light-show-breaks-guinness-world-records-title-olympic-winter-games-pyeongchang-2018/` 2018. [Online; accessed 30-September-2019].

[311] Craig W Reynolds. Flocks, herds and schools: A distributed behavioral model. In *Proceedings of the 14th annual conference on Computer graphics and interactive techniques*, pages 25–34, 1987.

[312] Fanny Riedo, Morgane Chevalier, Stéphane Magnenat, and Francesco Mondada. Thymio II, a Robot that Grows Wiser with Children. In *Proceedings of the IEEE Workshop on Advanced Robotics and its Social Impacts*, pages 187–193, 2013.

[313] James F. Roberts, Timothy S. Stirling, Jean-Christophe Zufferey, and Dario Floreano. Quadrotor Using Minimal Sensing For Autonomous Indoor Flight. In *Proceedings of the European Micro Air Vehicle Conference and Flight Competition*, 2007.

[314] Rui P Rocha, David Portugal, Micael Couceiro, Filipe Araújo, Paulo Menezes, and Jorge Lobo. The CHOPIN project: Cooperation between human and robotic teams in catastrophic incidents. In *Proceedings of the IEEE International Symposium on Safety, Security, and Rescue Robotics*, pages 1–4, 2013.

[315] Eric Rohmer, Surya PN Singh, and Marc Freese. V-rep: A versatile and scalable robot simulation framework. In *2013 IEEE/RSJ International Conference on Intelligent Robots and Systems*, pages 1321–1326. IEEE, 2013.

[316] Michael Rubenstein, Christian Ahler, Nick Hoff, Adrian Cabrera, and Radhika Nagpal. Kilobot: A low cost Robot with Scalable Operations Designed for Collective Behaviors. *Robotics and Autonomous Systems (Elsevier)*, 62(7):966–975, 2014.

[317] Michael Rubenstein, Christian Ahler, and Radhika Nagpal. Kilobot: A low cost scalable robot system for collective behaviors. In *2012 IEEE International Conference on Robotics and Automation*, pages 3293–3298, May 2012.

[318] Ray A. Russell. Ant trails - an example for robots to follow? In *Proceedings 1999 IEEE International Conference on Robotics and Automation (Cat. No.99CH36288C)*, volume 4, pages 2698–2703 vol.4, May 1999.

[319] Joan Saez-Pons, Lyuba Alboul, Jacques Penders, and Leo Nomdedeu. Multi-robot team formation control in the GUARDIANS project. *Industrial Robot: An International Journal*, 37(4):372–383, 2010.

[320] Erol Şahin. Swarm robotics: From sources of inspiration to domains of application. In *Proceedings of the International Workshop on Swarm Robotics*, pages 10–20, 2004.

[321] Muhammad Salman, David Garzón Ramos, Ken Hasselmann, and Mauro Birattari. Phormica: Photochromic pheromone release and detection system for stigmergic coordination in robot swarms. *Frontiers in Robotics and AI*, 7:195, 2020.

[322] William H. Sandholm. *Population Games and Evolutionary Dynamics*. MIT Press, 2010.

[323] Hiroki Sayama. Morphologies of self-organizing swarms in 3d swarm chemistry. In *Proceedings of the 14th Annual Conference on Genetic and Evolutionary Computation*, pages 577–584. ACM, 2012.

[324] Hiroki Sayama. PyCX: A python-based simulation code repository for complex systems education. *Complex Adaptive Systems Modeling*, 1(1):1–10, 2013.

[325] Jürgen Scherer, Saeed Yahyanejad, Samira Hayat, Evsen Yanmaz, Torsten Andre, Asif Khan, Vladimir Vukadinovic, Christian Bettstetter, Hermann Hellwagner, and Bernhard Rinner. An autonomous multi-uav system for search and rescue. In *Proceedings of the First Workshop on Micro Aerial Vehicle Networks, Systems, and Applications for Civilian Use*, pages 33–38, 2015.

[326] Thomas Schmickl, Stjepan Bogdan, Luís Correia, Serge Kernbach, Francesco Mondada, Michael Bodi, Alexey Gribovskiy, Sibylle Hahshold, Damjan Miklic, Martina Szopek, et al. Assisi: Mixing animals with robots in a hybrid society. In *Proceedings of the Conference on Biomimetic and Biohybrid Systems*, pages 441–443, 2013.

[327] Thomas Schmickl and Heiko Hamann. Beeclust: A swarm algorithm derived from honeybees. In Yang Xiao, editor, *Bio-inspired Computing and Networking*, pages 95–137. CRC Press, 2011.

[328] Thomas Schmickl, Martin Stefanec, and Karl Crailsheim. How a life-like system emerges from a simplistic particle motion law. *Scientific reports*, 6(1):1–15, 2016.

[329] Thomas Schmickl, Ronald Thenius, Christoph Möslinger, Gerald Radspieler, Serge Kernbach, and Karl Crailsheim. Get in touch: Cooperative decision making based on robot-to-robot collisions. *Autonomous Agents and Multi-Agent Systems*, 18(1):133–155, 2008.

[330] Thomas Schmickl, Ronald Thenius, Christoph Moslinger, Jon Timmis, Andy Tyrrell, Mark Read, James Hilder, Jose Halloy, Alexandre Campo, Cesare Stefanini, et al. Cocoro - the self-aware underwater swarm. In *Proceedings of the 5th IEEE Conference on Self-Adaptive and Self-Organizing Systems Workshops*, pages 120–126, 2011.

[331] Thomas Schmickl, Franz Wotawa, Ronald Thenius, and Joshua Cherian Varughese. Fstaxis algorithm: Bio-inspired emergent gradient taxis. In *Proceedings of the Artificial Life Conference 2016 13*, pages 330–337. MIT Press, 2016.

[332] Melanie Schranz, Alessandra Bagnato, Etienne Brosse, and Wilfried Elemenreich. Modelling a CPS swarm system: A simple case study. In *Proceedings of the International Conference on Model-Driven Engineering and Software Development (MODELSWARD)*, pages 615–624. SciTePress, January 2018.

[333] Melanie Schranz, Gianni A Di Caro, Thomas Schmickl, Wilfried Elmenreich, Farshad Arvin, Ahmet Şekercioğlu, and Micha Sende. Swarm intelligence and cyber-physical systems: Concepts, challenges and future trends. *Swarm and Evolutionary Computation*, 60:100762, 2020.

[334] Melanie Schranz, Kseniia Harshina, Peter Forgacs, and Fred Buining. Agent-based modeling in the edge continuum using swarm intelligence. In *Proceedings of the 16th International Conference on Agents and Artificial Intelligence*. SciTePress, 2024.

[335] Melanie Schranz, Micha Sende, Alessandra Bagnato, and Etienne Brosse. Modeling swarm intelligence algorithms for cps swarms. *Ada User Journal*, 40(3):169–177, September 2019.

[336] Melanie Schranz, Martina Umlauft, and Wilfriedi Elmenreich. Bottom-up job shop scheduling with swarm intelligence in large production plants. In *Proceedings of the 11th International Conference on Simulation and Modeling Methodologies, Technologies and Applications - SIMULTECH*. SciTePress, July 2021.

[337] Melanie Schranz, Martina Umlauft, Micha Sende, and Wilfried Elmenreich. Swarm robotic behaviors and current applications. *Frontiers in Robotics and AI*, 7:36, 2020.

[338] Adam M. Schroeder and Manish Kumar. Design of decentralized chemotactic control law for area coverage using swarm of mobile robots. In *Proceedings of the American Control Conference*, pages 4317–4322, 2016.

[339] Israel's Homeland Security. Fish swarm – model for energy-saving autonomous vehicle swarm. `https://i-hls.com/archives/84280`, 2018. [Online; accessed 30-September-2019].

[340] Madhubhashi Senanayake, Ilankaikone Senthooran, Jan Carlo Barca, Hoam Chung, Joarder Kamruzzaman, and Manzur Murshed. Search and tracking algorithms for swarms of robots: A survey. *Robotics and Autonomous Systems*, 75:422–434, 2016.

[341] Micha Sende. swarm_behaviors. `https://wiki.ros.org/swarm_behaviors`, 2019. [Online; accessed 23-March-2023].

[342] Micha Sende. swarm_functions. `https://wiki.ros.org/swarm_functions`, 2019. [Online; accessed 23-March-2023].

[343] Micha Sende. CPSwarm Hardware Functions. `https://github.com/cpswarm/hardware_functions`, 2023. [Online; accessed 23-March-2023].

[344] Micha Sende. Sensing and Actuation. `https://github.com/cpswarm/sensing_actuation`, 2023. [Online; accessed 23-March-2023].

[345] Micha Sende, Melanie Schranz, , Gianluca Prato, Etienne Brosse, Omar Morando, and Martina Umlauft. Engineering swarms of cyber-physical systems with the cpswarm workbench. *Journal of Intelligent & Robotic Systems*, 102(83):18, 2021.

[346] Sebastian Senge and Horst F Wedde. Bee-inpired road traffic control as an example of swarm intelligence in cyber-physical systems. In *Proceedings of the 38th Euromicro Conference on Software Engineering and Advanced Applications*, pages 258–265. IEEE, 2012.

[347] Jörg Seyfried, Marc Szymanski, Natalie Bender, Ramon Estaña, Michael Thiel, and Heinz Wörn. The I-SWARM Project: Intelligent Small World Autonomous Robots for Micro-manipulation. In *Proceedings of the International Workshop on Swarm Robotics*, pages 70–83, 2004.

[348] Shital Shah, Debadeepta Dey, Chris Lovett, and Ashish Kapoor. Airsim: High-fidelity visual and physical simulation for autonomous vehicles. In *Field and Service Robotics*, 2017.

[349] Hamed Shah-Hosseini. Principal components analysis by the galaxy-based search algorithm: A novel metaheuristic for continuous optimisation. *International Journal of Computational Science and Engineering*, 6(1-2):132–140, 2011.

[350] Adi Shklarsh, Gil Ariel, Elad Schneidman, and Eshel Ben-Jacob. Smart swarms of bacteria-inspired agents with performance adaptable interactions. *PLoS Computational Biology*, 7(9):1–11, 2011.

[351] Yoav Shoham and Kevin Leyton-Brown. *Multiagent systems: algorithmic, game-theoretic, and logical foundations.* Cambridge University Press, 2008.

[352] Ola Shorinwa, Javier Yu, Trevor Halsted, Alex Koufos, and Mac Schwager. Distributed multi-target tracking for autonomous vehicle fleets. In *2020 IEEE International Conference on Robotics and Automation (ICRA)*, pages 3495–3501. IEEE, 2020.

[353] Abdul A Siddiqi and Simon M Lucas. A comparison of matrix rewriting versus direct encoding for evolving neural networks. In *1998 IEEE International Conference on Evolutionary Computation Proceedings. IEEE World Congress on Computational Intelligence (Cat. No. 98TH8360)*, pages 392–397. IEEE, 1998.

[354] Jennifer Simonjan, Stefano Ricardo Probst, and Melanie Schranz. Inducing defenders to mislead an attacking uav swarm. In *Proceedings of the 42nd IEEE International Conference on Distributed Computing Systems Workshops*, pages 278–283, 2022.

[355] Jorge M Soares, Inaki Navarro, and Alcherio Martinoli. The Khepera IV Mobile Robot: Performance Evaluation, Sensory Data and Software Toolbox. In *Proceedings of the 2nd Iberian Robotics Conference*, pages 767–781, 2016.

[356] Anita Sobe. *Self-organizing Multimedia Delivery.* Phd thesis, Alpen-Adria Universität Klagenfurt, 2012.

[357] Anita Sobe and Wilfried Elmenreich. Smart microgrids: Overview and outlook. In *Proceedings of the ITG INFORMATIK Workshop on Smart Grids*, Braunschweig, Germany, September 2012.

[358] Anita Sobe, Wilfried Elmenreich, Tibor Szkaliczki, and Laszlo Böszörmenyi. SEAHORSE: Generalizing an artificial hormone system algorithm to a middleware for search and delivery of information units. *Computer Networks*, 80:124–142, 2015.

[359] Anita Sobe, Istvan Fehérvári, and Wilfried Elmenreich. FREVO: A tool for evolving and evaluating self-organizing systems. In *Proceedings of the 1st International Workshop on Evaluation for Self-Adaptive and Self-Organizing Systems*, pages 105–110, 2012.

[360] Ricard V. Solé, Susanna C. Manrubia, Bartolo Luque, Jordi Delgado, and Jordi Bascompte. Phase transitions and complex systems: Simple, nonlinear models capture complex systems at the edge of chaos. *Complexity*, 1(4):13–26, 1996.

[361] Kian Lun Soon, Joanne Mun-Yee Lim, and Rajendran Parthiban. Coordinated traffic light control in cooperative green vehicle routing for pheromone-based multi-agent systems. *Applied Soft Computing*, 81:105486, 2019.

[362] Kenneth Stanley. rtNEAT C++. `https://nn.cs.utexas.edu/?rtNEAT`, 2006. [Online; accessed 23-March-2023].

[363] Kenneth O. Stanley and Risto Miikkulainen. Evolving Neural Networks through Augmenting Topologies. *Evolutionary Computation*, 10(2):99–127, 2002.

[364] David Steber, Peter Bazan, and Reinhard German. Swarm - increasing households' internal pv consumption and offering primary control power with distributed batteries. In *Lecture Notes in Computer Science (including subseries Lecture Notes in Artificial Intelligence and Lecture Notes in Bioinformatics)*, volume 9424, pages 3–11. Springer Verlag, 2015.

[365] Nathalie Steinhauer, Kelly Kulhanek, Karina Antúnez, Hannelie Human, Panuwan Chantawannakul, Marie-Pierre Chauzat, and Dennis van Engelsdorp. Drivers of colony losses. *Current Opinion in Insect Science*, 26:142 – 148, 2018.

[366] John D Sterman. System dynamics modeling: tools for learning in a complex world. *California management review*, 43(4):8–25, 2001.

[367] Apache Storm. `https://storm.apache.org/`. [Online; accessed 8-August-2024].

[368] Daniel P Stormont. Robot swarms for planetary exploration. In *Proceedings of the 4th International Conference and Exposition on Robotics for Challenging Situations and Environments*, pages 347–352, 2000.

[369] Steven H. Strogatz. *Nonlinear Dynamics and Chaos: With Applications to Physics, Biology, Chemistry, and Engineering.* Studies in Nonlinearity. Avalon Publishing, 2014.

[370] Daniel Strömbom, Mattias Siljestam, Jinha Park, and David JT Sumpter. The shape and dynamics of local attraction. *The European Physical Journal Special Topics*, 224(17):3311–3323, 2015.

[371] Gerald Jay Sussman. Building robust systems an essay. *Massachusetts Institute of Technology*, 2007.

[372] Project SWARMIX: Synergistic interactions in swarms of heterogeneous agents (2011-2014). `http://www.swarmix.org/`. [Online; accessed 14-August-2019].

[373] S. Swetha, R. Anandan, and K. Kalaivani. An Investigation of Micro Aerial Vehicles (μAV). *International Journal of Engineering & Technology*, 7(2.31):174–177, 2018.

[374] Martina Szopek, Thomas Schmickl, Ronald Thenius, Gerald Radspieler, and Karl Crailsheim. Dynamics of collective decision making of honeybees in complex temperature fields. *PloS one*, 8(10):e76250, 2013.

[375] Kenichi Tamura and Keiichiro Yasuda. Spiral dynamics inspired optimization. *Journal of Advanced Computational Intelligence and Intelligent Informatics*, 15(8):1116–1122, 2011.

[376] Ying Tan and Yuanchun Zhu. Fireworks algorithm for optimization. In *Proceedings of the International Conference on Swarm Intelligence*, pages 355–364. Springer, 2010.

[377] Arash Tavakoli, Haig Nalbandian, and Nora Ayanian. Crowdsourced coordination through online games. In *Proceedings of the 11th ACM/IEEE International Conference on Human-Robot Interaction*, pages 527–528, 2016.

[378] TensorFlow. `https://www.tensorflow.org/`. [Online; accessed 8-August-2024].

[379] Erich C. Teppan. Dispatching rules revisited – a large scale job shop scheduling experiment. In *Proceedings of the IEEE Symposium Series on Computational Intelligence*, pages 561–568, 2018.

[380] Ronald Thenius, Daniel Moser, Joshua Cherian Varughese, Serge Kernbach, Igor Kuksin, Olga Kernbach, Elena Kuksina, Nikola Mišković, Stjepan Bogdan, Tamara Petrović, et al. subCULTron-Cultural Development as a Tool in Underwater Robotics. In *Proceedings of the Artificial Life and Intelligent Agents Symposium*, pages 27–41. Springer, 2016.

[381] Jan C Thiele. R marries NetLogo: introduction to the RNetLogo package. *Journal of Statistical Software*, 58:1–41, 2014.

[382] Seth Tisue and Uri Wilensky. Netlogo: A simple environment for modeling complexity. In *International conference on complex systems*, volume 21, pages 16–21. Boston, MA, 2004.

[383] Ioan C. Trelea. The particle swarm optimization algorithm: Convergence analysis and parameter selection. *Information Processing Letters*, 85(6):317–325, 2003.

[384] Elio Tuci and Alexandre Rabérin. On the design of generalist strategies for swarms of simulated robots engaged in a task-allocation scenario. *Swarm Intelligence*, 9(4):267–290, 2015.

[385] Ali E Turgut, Hande Çelikkanat, Fatih Gökçe, and Erol Şahin. Self-organized flocking in mobile robot swarms. *Swarm Intelligence*, 2:97–120, 2008.

[386] Ali Emre Turgut, İhsan Caner Boz, İlkin Ege Okay, Eliseo Ferrante, and Cristián Huepe. Interaction network effects on position-and velocity-based models of collective motion. *Journal of the Royal Society Interface*, 17(169):20200165, 2020.

[387] Alan M. Turing. The chemical basis of morphogenesis. *Philosophical Transactions of the Royal Society of London. Series B. Biological Sciences*, 237(641):37–72, 1952.

[388] Fritz Ulbrich, Simon Sebastian Rotter, and Raul Rojas. Adapting to the traffic swarm: Swarm behaviour for autonomous cars. In Ying Tan, editor, *Handbook of Research on Design, Control, and Modeling of Swarm Robotics*, chapter 10, pages 263–285. IGI Global, 2016.

[389] Martina Umlauft. Swarm Fab Simulation. `https://swarmfabsim.github.io`, 2023. [Online; accessed 23-March-2023].

[390] Martina Umlauft, Melanie Schranz, and Wilfried Elmenreich. Swarm-FabSim: A simulation framework for bottom-up optimization in flexible job-shop scheduling using Netlogo. In *Proceedings of the 12th International Conference on Simulation and Modeling Methodologies, Technologies and Applications - SIMULTECH*, pages 271–279. SciTePress, July 2022.

[391] Gabriele Valentini, Anthony Antoun, Marco Trabattoni, Bernát Wiandt, Yasumasa Tamura, Etienne Hocquard, Vito Trianni, and Marco Dorigo. Kilogrid: a novel experimental environment for the kilobot robot. *Swarm Intelligence*, 12(3):245–266, Sep 2018.

[392] Richard Vaughan. Massively multi-robot simulation in stage. *Swarm intelligence*, 2(2-4):189–208, 2008.

[393] Ganesh K Venayagamoorthy. Potentials and promises of computational intelligence for smart grids. In *2009 IEEE Power & Energy Society General Meeting*, pages 1–6. IEEE, 2009.

[394] Tamás Vicsek, András Czirók, Eshel Ben-Jacob, Inon Cohen, and Ofer Shochet. Novel type of phase transition in a system of self-driven particles. *Physical Review Letters*, 75(6):1226–1229, 1995.

[395] Tamás Vicsek and Anna Zafeiris. Collective motion. *Physics reports*, 517(3-4):71–140, 2012.

[396] 4D Virtualiz. `https://www.4d-virtualiz.com/en/`, 2024. [Online; accessed 23-March-2023].

[397] Alessandra Vitanza, Paolo Rossetti, Francesco Mondada, and Vito Trianni. Robot Swarms as an Educational Tool: The Thymio's Way. *International Journal of Advanced Robotic Systems*, 16(1), 2019.

[398] Richard G Vogt, Lynn M Riddiford, and Glenn D Prestwich. Kinetic properties of a sex pheromone-degrading enzyme: the sensillar esterase of antheraea polyphemus. *Proceedings of the National Academy of Sciences*, 82(24):8827–8831, 1985.

[399] Mostafa Wahby, Mary Katherine Heinrich, Daniel Nicolas Hofstadler, Ewald Neufeld, Igor Kuksin, Payam Zahadat, Thomas Schmickl, Phil Ayres, and Heiko Hamann. Autonomously shaping natural climbing plants: A bio-hybrid approach. *Open Science*, 5(10):180296, 2018.

[400] Pangwei Wang, Hui Deng, Juan Zhang, Li Wang, Mingfang Zhang, and Yongfu Li. Model predictive control for connected vehicle platoon under switching communication topology. *IEEE Transactions on Intelligent Transportation Systems*, 2021.

[401] Shiyi Wang, Ali E Turgut, Thomas Schmickl, Barry Lennox, and Farshad Arvin. Investigation of cue-based aggregation behaviour in complex environments. In *International Conference on Collaborative Computing: Networking, Applications and Worksharing*, pages 18–36. Springer, 2020.

[402] Nicholas W. Watkins, Gunnar Pruessner, Sandra C. Chapman, Norma B. Crosby, and Henrik J. Jensen. 25 years of self-organized criticality: Concepts and controversies. *Space Science Reviews*, 198(1):3–44, 2016.

[403] Hongxing Wei, Yingpeng Cai, Haiyuan Li, Dezhong Li, and Tianmiao Wang. Sambot: A Self-assembly Modular Robot for Swarm Robot. In *Proceedings of the IEEE International Conference on Robotics and Automation*, pages 66–71. IEEE, 2010.

[404] Uri Wilensky. Netlogo, 1999. `http://ccl.northwestern.edu/netlogo/`.

[405] Sean Wilson, Aurélie Buffin, Stephen C Pratt, and Spring Berman. Multi-robot Replication of Ant Collective Towing Behaviours. *Royal Society open science*, 5(10):180409, 2018.

[406] Sean Wilson, Ruben Gameros, Michael Sheely, Matthew Lin, Kathryn Dover, Robert Gevorkyan, Matt Haberland, Andrea Bertozzi, and Spring Berman. Pheeno, a Versatile Swarm robotic Research and Education Platform. *IEEE Robotics and Automation Letters*, 1(2):884–891, 2016.

[407] Wayne H. Wolf. Hardware-software co-design of embedded systems. *Proceedings of the IEEE*, 82(7):967–989, 1994.

[408] Kefan Wu, Junyan Hu, Zhengtao Ding, and Farshad Arvin. Distributed bearing-only formation control for heterogeneous nonlinear multi-robot systems. *IFAC-PapersOnLine*, 56(2):3447–3452, 2023.

[409] Kefan Wu, Junyan Hu, Zhengtao Ding, and Farshad Arvin. Finite-time fault-tolerant formation control for distributed multi-vehicle networks with bearing measurements. *IEEE Transactions on Automation Science and Engineering*, 2023.

[410] Kefan Wu, Junyan Hu, Zhenhong Li, Zhengtao Ding, and Farshad Arvin. Distributed collision-free bearing coordination of multi-uav systems with actuator faults and time delays. *IEEE Transactions on Intelligent Transportation Systems*, 2024.

[411] Tristram D. Wyatt. *Pheromones and Animal Behavior: Chemical Signals and Signatures.* Cambridge University Press, 2014.

[412] Wyss Institute. Programmable Robot Swarms. `https://wyss.harvard.edu/technology/programmable-robot-swarms/`, 2017. [Online; accessed 06-November-2019].

[413] Songtao Xie, Junyan Hu, Parijat Bhowmick, Zhengtao Ding, and Farshad Arvin. Distributed motion planning for safe autonomous vehicle overtaking via artificial potential field. *IEEE Transactions on Intelligent Transportation Systems*, 23(11):21531–21547, 2022.

[414] Songtao Xie, Junyan Hu, Zhengtao Ding, and Farshad Arvin. Collaborative overtaking of multi-vehicle systems in dynamic environments: A distributed artificial potential field approach. In *2021 20th International Conference on Advanced Robotics (ICAR)*, pages 873–878. IEEE, 2021.

[415] Songtao Xie, Junyan Hu, Zhengtao Ding, and Farshad Arvin. Cooperative adaptive cruise control for connected autonomous vehicles using spring damping energy model. *IEEE Transactions on Vehicular Technology*, 2022.

[416] Xin-She Yang. Flower pollination algorithm for global optimization. In *Proceedings of the International Conference on Unconventional Computing and Natural Computation*, pages 240–249, 2012.

[417] Xin-She Yang, Zhihua Cui, Renbin Xiao, Amir Hossein Gandomi, and Mehmet Karamanoglu. *Swarm Intelligence and Bio-inspired Computation: Theory and Applications*. Elsevier, 2013.

[418] Xin-She Yang, Suash Deb, Yu-Xin Zhao, Simon Fong, and Xingshi He. Swarm intelligence: Past, present and future. *Soft Computing*, pages 1–11, 2017.

[419] Shoji Yano. Kajima to develop automated construction machinery for building on Mars, Moon. `https://asia.nikkei.com/Tech-Science/Tech/Kajima-to-develop-automated-construction-machinery-for-building-on-Mars-moon`, 2016. [Online; accessed 9-July-2019].

[420] Sha Yi, Zeynep Temel, and Katia Sycara. Puzzlebots: Physical coupling of robot swarms. In *2021 IEEE International Conference on Robotics and Automation (ICRA)*, pages 8742–8748, 2021.

[421] Bin Yu, Zhong-Zhen Yang, and Baozhen Yao. An improved ant colony optimization for vehicle routing problem. *European Journal of Operational Research*, 196(1):171–176, 2009.

[422] Zhi Yuan, Marco A. Montes de Oca, Mauro Birattari, and Thomas Stützle. Continuous optimization algorithms for tuning real and integer parameters of swarm intelligence algorithms. *Swarm Intelligence*, 6(1):49–75, 2012.

[423] Xizhe Zang, Sajid Iqbal, Yanhe Zhu, Xinyu Liu, and Jie Zhao. Applications of chaotic dynamics in robotics. *International Journal of Advanced Robotic Systems*, 13(2), 60 p., 2016.

[424] Fangbo Zhang, Andrea L. Bertozzi, Karthik Elamvazhuthi, and Spring Berman. Performance bounds on spatial coverage tasks by stochastic robotic swarms. *IEEE Transactions on Automatic Control*, 63(6):1473–1488, 2018.

[425] Guohui Zhang, Xinyu Shao, Peigen Li, and Liang Gao. An effective hybrid particle swarm optimization algorithm for multi-objective flexible job-shop scheduling problem. *Computers & Industrial Engineering*, 56(4):1309–1318, 2009.

[426] Yuan Zhang, Yu Yuan, and Kejing Lu. E-commerce information system data analytics by advanced aco for asymmetric capacitated vehicle delivery routing. *Information Systems and e-Business Management*, pages 1–19, 2019.

[427] Yating Zheng, Cristián Huepe, and Zhangang Han. Experimental capabilities and limitations of a position-based control algorithm for swarm robotics. *Adaptive Behavior*, 30(1):19–35, 2022.

[428] George R. Zug. Walk pattern analysis of cryptodiran turtle gaits. *Animal Behaviour*, 20(3):439–443, 1972.

[429] Victor Zykov, Efstathios Mytilinaios, Mark Desnoyer, and Hod Lipson. Evolved and Designed Self-Reproducing Modular Robotics. *IEEE Transactions on Robotics*, 23(2):308–319, 2007.

Index

T

Introduction

Being a member of a historically marginalized social group is risky for your health, especially in the United States, where inequality is high, and those pushed to the margins have worse health outcomes than their more advantaged counterparts, despite relatively high standards of living as compared with some regions of the world. For more than sixty years, the federal community health center program, along with Medicaid, has been the federal government's primary response to addressing the issue of race and income-based health disparities in the United States. While biology and genetics are essential precursors to the variation in health outcomes for individuals, social conditions are also fundamentally important as they disproportionately influence individuals' exposures to health risks and protective health factors. Community health centers focus on neutralizing the effects of social barriers to care by providing high-quality, low-cost primary care services in low-income communities with limited access to healthcare. With more than 1,400 centers providing care through 14,000 service sites nationwide, community health centers are a lifeline for uninsured and underinsured families and individuals, providing critical access to vaccines, preventative healthcare, and health-related social services such as nutrition and insurance enrollment. In 2021, community health centers provided care to over 30 million patients, including 1.3 million patients experiencing homelessness, equivalent to one-third of all Americans experiencing poverty (National Association of Community Health Centers [NACHC], 2023). While some of these patients are uninsured, most have Medicaid or private insurance that does not sufficiently cover all of the patient's health needs (e.g., family planning and contraceptive services).

Community health centers receive significant subsidies from the federal and state governments to offset the financial burden of predominantly servicing Medicaid, uninsured, and underinsured patient populations. This financial support enables them to provide quality primary healthcare without the concern of profitability. Nonetheless, efficiency, performance, and value are significant concerns for external stakeholders of the program who entrust community health centers with public funds to provide a cost-effective medical safety net to those most in need. Securing an annual appropriation of over $5.7 billion, community health centers have enjoyed positive public sentiment and bipartisan support since their establishment as a War on Poverty program in 1965 and have never been defunded or discontinued—a rarity for a social policy program in the United States. Initially, the centers and their authorized program were positioned as an effective means of addressing the health-related causes and consequences of poverty in the United States during a time when poverty eradication was at the forefront of domestic policy. As a prominent economics expert noted in 1964, "No one is in favor of poverty, and criticism of a program purporting to deal with it is as risky politically as being in favor of sin or against motherhood" (Hunter, 1964, p. E5). This sociopolitical climate, primarily energized by the desire to keep the recently assassinated JFK's presidential agenda alive, provided fertile ground for establishing a growing federal effort to address indigent healthcare in the United States. However, over time, public opinion and presidential agendas changed, which endangered all of the social programs of the 1960s. However, despite these challenges, the community health center program has been a standout example of how program administrators and advocates can keep pace, performing an intricate dance of staying true to the programmatic mission while aligning perceptions of the program with fluctuating public and legislative opinion on indigent healthcare in the United States. The traditional media have been an indispensable partner in this effort, illuminating the stories of community health centers and their patients and matching them with contemporary definitions and contexts of healthcare, poverty, and the government's responsibility to address both through policy efforts.

The public's reaction to social health crises is often shaped by media coverage, which historically oscillates between apathy and a strong, albeit fleeting, collective response in times of crisis (Duffy, 1992). This underscores the crucial role of the media in shaping public perception and the need for accurate and responsible reporting. Federal public health initiatives have generally received sporadic attention from legislators throughout history due

to the public's varying levels of concern and interest. Although initially intense and all-encompassing, Covid-19 was no exception to this pattern. The nation's initial collective response to the pandemic was rapt attention and compliance, but began to wane and devolve into protest as stay-at-home and face-covering orders became economically, politically, and socially cumbersome (Bosman et al., 2020). Even more concerning is the public's tendency to perceive public health efforts as merely targeting deviant behaviors of a small subset of the population who are sick due to their own poor choices and behaviors. The public shaming of Black and Hispanic communities facing the brunt of the Covid-19 pandemic is a good example of this (Kendi, 2020). We can also contemporarily look at the public response to the HIV/AIDS epidemic in the 1980s (Clarke et al., 2006) or even media coverage of the opioid versus heroin epidemics (Netherland & Hansen, 2016). The problem of stigma is exacerbated by the fact that medically vulnerable populations are disproportionately comprised of those belonging to historically marginalized social groups and other populations pushed to the margins of society. Because these groups hold less social and political power, legislators can treat them as an afterthought with little consideration from the federal government or the public (Kant & Rushefsky, 2005). Stigmatized responses to public health threats can translate into less robust media coverage as the perceived risk shifts from collective to individual, ultimately resulting in lower levels of congressional support as the public moves on to more pressing concerns.

Recognizing the distinct health needs of historically marginalized populations and communities, the community health center program strives to provide a broader and more culturally sensitive scope of quality healthcare than traditional healthcare providers. By design, centers are located in neighborhoods with high poverty rates and limited access to healthcare providers. Describing their care as patient-centered, the centers seek to treat the whole person's needs instead of solely focusing on diagnosing and treating disease. Empirical studies of community health center efficacy have highlighted various benefits of the program, including a reduction in racial disparities for preventive health measures such as early prenatal care, cancer screening, and eye and foot care for people with diabetes (Shi et al., 2004, 2009), and improved access to care for uninsured and Medicaid community health center patients as compared to uninsured and Medicaid patients receiving care at noncommunity health center facilities (Shi et al., 2007). While efficacy studies demonstrate a positive association between community health center utilization and health status for those traditionally facing barriers to care, there is a gap in the literature on whether community health center availability benefits an individual even when the center is not the individual's primary source of care. This distinction is vital to the story of community health centers as the primary message of the program is access, and far more individuals have access to community health centers than are regular users. Therefore, it is critical to understand how community health center availability affects the health of individuals regardless of whether they are regular users of the services and how social and structural factors impacting the individual affect this primary relationship.

As federal mandates require, centers with a federally qualified community health center designation provide primary health services such as physicals, urgent care, diagnostic testing, and assistance managing chronic disease regardless of an individual's ability to pay. They are also strategically located in urban, rural, and suburban areas with few options for quality healthcare, significantly increasing access to medical providers and services for those living in those areas. Auxiliary services such as dental, mental and behavioral health, and nutrition are also commonly offered by agencies with the designation, and all services are provided in a culturally competent manner to reduce social barriers to care. Community health center administrators are deeply committed to the program and communities and are exemplary examples of responsible public administrators. As a profession, they have faced and weathered many challenges to the program's existence over the years and, with the help of key advocates and legislators, kept the program alive and thriving.

As a War on Poverty demonstration program in its earliest days, the community health center program was one of many social initiatives formally established as a federal effort during President Johnson's administration to support the government's comprehensive approach to ameliorating poverty in the 1960s. As such, providing primary healthcare through community health centers was a means of empowering individuals and neighborhoods so that individuals from historically marginalized communities could fu participate in a bevy of new social initiatives designed to lift people out of the experience of poverty. For instance, social program administrators found it was nearly impossible for teenagers to fully participate in a job training program if they had undiagnosed migraines that left them debilitated during the day. Having access to a physician who can not only diagnose their condition but also prescribe low-cost medications to treat the symptoms and then refer the patient to an in-house counselor to provide some mental health support and behavioral techniques for coping with migraines not only addresses the health concern but also puts the young adult in a better position to flourish in society.

By serving as a means to the end of poverty eradication, early community health centers were deeply embedded and trusted neighborhood partners. Today, the program's mission remains the same, but the public's perception of it and its image have changed. Community health centers are now positioned as a medical safety net, catching Medicaid-eligible or uninsured individuals before they make the costly decision to forego preventive medical care or turn to emergency rooms for routine issues. These undesirable choices are expensive in the long run for individuals and the healthcare system as a whole, which elected officials and the public seek to avoid. The administrators and advocates of the program purport that the program primarily benefits the nation as a money-saver, reducing the overall costs of indigent healthcare to society. These cost savings and the program's willingness to absorb the patients that traditional medical providers prefer not to treat undergird the program's bipartisan success for the past six decades.

While the community health center program is an administrative success story, there is an unintended consequence of the program's heavy reliance on advocacy and legislative support. A narrative shift from being a place of personal and neighborhood empowerment to being a prong in an intersectoral system for fortifying the healthcare structure has weakened the power of community health centers to make transformational improvements in health disparities on a community level. One direct implication of this transition from empowerment to access is the diminished emphasis on identifying and mitigating fundamental causes of health disparities in favor of a more general one-size-fits-all approach to the distribution of care. The danger of the latter proach is that the effect of access on an individual's health will likely differ based on their social characteristics and the social context in which they live. Therefore, by engaging a unidimensional solution to the problem of disparities, the problem may actually be exacerbated for those with disproportionate barriers to care, as those with medical and social needs beyond basic access to healthcare are pushed to the margins of the healthcare safety net system.

About This Book

Adapting to the sociopolitical environment while upholding the program's mission is a crucial survival skill for contemporary administrators leading public organizations. Successful administrators often enjoy long-term, bipartisan support for their programs, robust budgets, and consistent congressional reauthorizations. This balance can be particularly challenging for administrators of federally subsidized public programs, who must manage their organization's social image within the legislative environment while also managing public expectations. Administrative flexibility can lead to mission creep if not managed carefully.

For nearly six decades, community health center administrators and advocates have diligently ensured the sustainability of the community health center program. They have strategically managed the program's image using the media and critical legislative connections. The program's mission, rooted in community empowerment, has remained steadfast. However, its societal purpose has shifted. Once a beacon of community empowerment, the program's public image has transitioned to emphasizing individual responsibility. This shift, while a necessary administrative response to the funding environment, has inadvertently diminished the emphasis on community empowerment as a primary tool for addressing health disparities at a neighborhood level, leading to unintended consequences, such as the marginalization of individuals living in the target communities, but are not regular users of community health center services. This service gap is important because of the program's historical commitment to neighborhoods, not just the compliant individual; it is almost impossible to change community health when the unit of analysis is not the community.

Two Steps Forward, Three to the Side delves into the delicate balance that federally subsidized social agency administrators and their advocates must strike to maintain programmatic distinctiveness and survival, using the federal community health center program's narrative transition as an illustrative lens. Chronologically structured by the program's legislative history across federal reauthorization periods, this book employs a sociological lens in the content analysis of media coverage and legislative records to trace the transformation of community health center program narratives. The examination uncovers the gradual de-emphasis of community empowerment as the purpose and benefit of the program in favor of messages about programmatic efficiency and cost savings to the broader healthcare system in response to challenging funding environments. On the one hand, the book contends that this intentional administrative strategy can be celebrated as the program's federal allocation has ballooned over decades of relationship-building and image management with the media and legislators during federal reauthorization periods. On the other hand, the shift represents a warning about de-emphasizing social justice objectives as part of a strategy to appease more conservative stakeholders. Overall, this book argues that the public administrators' strategic efforts to align the community health center program's image with public opinion and legislative priorities have resulted in a shift away from the program's anti-poverty roots, thus negatively impacting its effectiveness in addressing the root cause of neighborhood-level health disparities. The most concerning consequence is the persistent health disparities in the communities that they serve. Building on existing calls for the program to return to its community revitalization and anti-poverty roots, this book aims to both celebrate and caution against administrative strategies that align localized organizational outcomes with macroeconomic standards of success.

The chapters of this book are organized around themes that build out the assumptions and evidence for the argument mentioned above. Chapter 1 provides a historical overview of the federal community health center program, tracing its beginnings and growth over the last 60 years. It also contextualizes the program within the relationship between health and poverty, helping to explain why the program is so critical in the United States. Chapter 2 discusses the complexity and challenges of the policymaking process in the US political process and how these challenges have shaped the program's efforts to remain viable on the federal landscape. Due to the openness of the democratic system to public opinion and macro-level pressures, astute administrators must develop a strategy to remain relevant and in good standing with the public and legislators. For community health centers, the media have been a key partner in the communication and co-creation of positive social narratives about the critical nature of their work. Chapter 3 contains the core analysis, tracing how the social image of community health centers and their patients have changed over several Presidential administrations. Thematic evidence from State of the Union Addresses, legislative documents, and *New York Times* articles provides an illustrative lens of how the program's image has kept lockstep with Presidential agendas and public sentiment about indigent healthcare. Chapter 4 examines the unintended consequences of the program's narrative shift, paradoxically highlighting the persistence of health disparities in the communities where health centers are located, illustrating the urgent need for reevaluating the program's priorities and a renewed focus on neighborhoods. Questions about measuring the success of public policy initiatives and the isolated effect of having access to a community health center are explored. Finally, the Conclusion challenges the reader to consider the community health center program's future purpose and role in the larger healthcare landscape. In it, I argue for a return to the roots of poverty eradication and neighborhood empowerment to address the root cause of health disparities more effectively and fundamentally.

The Dance Metaphor

This book uses dance as both a verb and a noun as a metaphor for the strategic administration process in public agencies. As a verb, it speaks to the movement and flexibility required to effectively serve all stakeholders impacted by the agency's work, most notably the target populations and funders. Competing stakeholder needs and expectations require administrators to be graceful while also under fire as they strive to keep their organizations viable, distinct, and funded. Stakeholders are rarely interested in witnessing the struggle and the pain associated with the performance of effective leadership, just as the audience does not want to see the battered and calloused feet of the dancer. Funders and the public, who are also part of this dance, are committed to the beauty of the change knowing that their resources and contributions have been masterfully spun into practical solutions for the social issues the problems seek to eradicate.

When used as a noun, dance can refer to a product, an event, or a place. As a product for consumption, a dance attracts observers seeking to connect with art, culture, storytelling, and emotional experience. Similarly, the products and outcomes of public service agencies fulfill a social desire for mainstream society to connect with a specific subculture, perhaps one to which they belong or that of another, to feel a connection and purpose. For clients of public agencies, the dance concept is more tangible, like the patron who comes to a place to dance for interpersonal connection, emotional release, and self-care. When dance is a place, there are rules of engagement and social expectations of quality, especially when money is exchanged for access. This transformative power of public service, akin to the transformative power of dance, inspires and drives the work of public administrators, shaping their vision, agility, and creativity.

Whatever way it is defined, dance is complicated and transformative. It creates meaning between the dancer and the environment that can be seen, touched, and heard. The art of public administration is also much like dance. Creating and cultivating a positive organizational image for public agencies requires vision, agility, and creativity. Leading in the public sector is subject to a level of scrutiny that no other sector endures. The dance is collaborative and impactful. Every move communicates meaning, whether assumed or explicit. Therefore, agency administrators must take care to master the imagery and messages they create. They must know their audience and be careful of pitfalls and sensitivities that can turn them away. As administrators manage resources on behalf of funders and the public, they must remember the routine and be careful not to misstep or fall. Moreover, when they reach the end of an initiative or project, they can take a brief bow. However, they must assess the performance and reset in preparation for the next engagement, as the administration is never-ending. This emphasis on communication ensures that all involved are informed and aware, fostering a sense of transparency and trust.

1

An Invitation to Engage

Establishing Community Health Centers' Place in the US Healthcare System

Community health centers and their provision of primary healthcare services to medically vulnerable populations did not formally exist in the mid-twentieth century. Primary medical care was mainly reserved for those who could afford the fees of private doctors making home visits, and public hospitals had limited beds for the acutely ill in danger of threatening public health. Charity care did exist through benevolent arms of medical and religious institutions, but was mostly for those deemed as worthy poor: good people facing health challenges through no fault or deviant actions of their own. Entire indigent communities, especially those predominantly comprised of racial and ethnic minorities, had rarely, if ever, seen a traditional doctor, and of those that had, did so under extreme hardship and sacrifice unless they were abnormally wealthy. Instead, home and cultural remedies of variable efficacy were the norm for these populations, and many went without any care at all, significantly impacting their social outcomes and mobility.

This chapter chronicles the growth of the community health center program, which started as an anti-poverty demonstration project with just two sites nationally in 1965. It has since evolved into a federally authorized program with more than 14,000 service sites today. The chapter begins with a description of the community health center program, its purpose, and the reasons for its establishment. It then delves into the broader issue of social factors as the fundamental cause of disease, a crucial concept for understanding the unique role of community health centers in combating poverty and poor health outcomes for medically vulnerable and historically marginalized populations. The chapter then traces the history of the community health center program, from the sociopolitical environment that gave rise to Johnson's War on Poverty initiative to the motivations, inspirations, and actions of the program's founders. It concludes with a description of the first two established centers in Mississippi and Massachusetts and the initial program requirements from the federal agency responsible for oversight.

What Exactly Is a Community Health Center?

Of the dozens of programs established during President Lyndon B. Johnson's War on Poverty effort, the community health center program is one of the few enduring success stories. Since its inception, the program has grown exponentially in terms of funding, patients seen, scope of services, and bipartisan support. While there have been some budget years where lobbying efforts were necessarily stronger than others, the program has never been defunded or discontinued. This rare continuity of a War on Poverty program reflects the successful efforts of advocates and administrators who never leave the program's fate to chance.

Demographically, community health center patients represent some of the most historically marginalized communities in the United States (NACHC, 2023). In 2021, community health centers served over 30 million patients, 67 percent of whom lived at or below the federal poverty level, equivalent to $26,500 for a family of four. With Medicaid expansion, 20 percent of community health center patients seen in 2021 were uninsured, and 48 percent were enrolled in Medicaid. Individuals identifying as racial and/or ethnic minorities were also overrepresented in the community health center patient population at 63 percent of all patients seen in the calendar year.

Two distinguishing characteristics of community health centers are (1) their locations in communities with high poverty rates and limited access to healthcare providers and (2) their commitment to providing high-quality, culturally competent primary healthcare services to anyone presenting for care, regardless of their ability to pay. Specializing in patient-centered, holistic care, community health centers seek to treat the needs of the whole person instead of solely focusing on diagnosing and treating disease. In addition to primary care services, community health centers offer mental health, nutritional, and social services, as well as enabling services such as transportation support and discounted medications. The centers also actively assist eligible uninsured patients with the insurance application process for Medicaid and Medicare coverage and coverages available through the Affordable Care Act to enroll as many uninsured patients in state-subsidized insurance plans as possible. Through patient navigators, many community health centers also assist with coordinating and scheduling specialist appointments for patients unfamiliar with hospital systems. The centers function in both urban and rural areas, and there are also centers with specialized designations that predominately focus on unhoused as well as migrant populations.

While firmly rooted in its altruistic mission to serve as a medical safety net for historically marginalized communities, there is also a substantial financial incentive for primary healthcare facilities to apply for and receive formal designation as a federally qualified community health center (FQHC). The status of FQHC, or FQHC Look-Alike for health centers that meet the requirements of the program but do not receive federal funding, confers several financial benefits to community health centers that are not available to traditional primary healthcare providers. Community health centers with FQHC status can apply for federal, state, and local grants to support capital and operational costs, receive discounted medications through the 340B program, access physicians through the National Health Service Corps (NHSC), and receive group medical malpractice liability protection at no cost under the Federal Torts Claims Act of 1992 and 1995.[1] The cost-based Medicaid payment structure is the most substantial financial benefit for community health centers. To subsidize uncompensated care and minimize the outlay of federal grants, states provide wraparound payments to community health centers at a rate equal to the centers' fully loaded cost of care, including operational expenses. This Medicaid reimbursement rate is significantly higher than that offered to private physicians and is in addition to a monthly capitation payment.

The low Medicaid reimbursement rates for private physicians create a disincentive for private physicians to service Medicaid patients while simultaneously incentivizing community health centers to grow their Medicaid base to meet demand. A 2011 survey of ambulatory primary care physicians revealed that although 96 percent of physicians accepted new patients, 31 percent were unwilling to accept new Medicaid patients (Decker, 2012). The author also found that physicians in metropolitan areas were less likely to accept Medicaid than physicians in more rural areas. Four years later, this finding was replicated through a similar study indicating that general practice and family physicians were far less likely to accept new Medicaid patients as compared to new private insurance or Medicare patients by a margin of more than 20 percentage points (Holgash & Heberlein, 2019). In contrast, community health centers seek partnerships with the state to become a preferred provider in the Medicaid primary care network. In the most successful partnerships, a community health center is the default primary care physician for new enrollees who do not specify a provider or plan at the time of enrollment.

Why Do We Need Community Health Centers?

The American healthcare system primarily consists of two models of care: curative, emphasizing the treatment and cure of disease, and preventative, emphasizing disease prevention. The curative model, characterized by cutting-edge technologies in the advanced treatment of disease, such as curing cancer or mapping human DNA, is the most lucrative aspect of the modern medical profession, as it tends to receive the most attention, funding, and consideration from legislators and the general public. As such, influential medical organizations, such as the American Medical Association (AMA), tend to focus their energies on the maintenance and growth of funding streams specific to this aspect of health, pushing the preventative model to the fringes of the profession and even into the realm of the public health sector, especially for historically marginalized populations. Conversely, public health initiatives have generally been associated with the preventative care model as characterized by the sanitation effort of the nineteenth century, the push for immunizations in the twentieth century, and the recent response to the Covid-19 pandemic.

Withstanding the Covid-19 response and the Affordable Care Act, federal public health responses have traditionally been relegated to background noise with spasms of widespread interest in the landscape of public policy debates. Historically, public reaction to social health crises in the United States has oscillated between apathy, such as the response to malaria and the flu, fear, like the initial response to Ebola and HIV, and a strong collective response, such as the response to yellow fever, smallpox, and Covid-19. These historical examples underscore the need for collective action in addressing health disparities. In their historical review of public health, Kant and Rushefsky (2005) argue that the public health sector has long suffered from "benign neglect" by the federal government. Moreover, public health efforts are often perceived as targeting deviant behaviors of a small subset of the population who are sick due to their own poor choices and behaviors, which has translated into low levels of support and funding in Congress. Because these target populations are perceived to be disproportionately minority and/or economically marginalized, the general population does not often recognize the sweeping positive effects of public health efforts for society as a whole, leaving most public health activities to be managed at the state and local level under block grants with little interference from the federal level.

So why do legislators care about the health of people living in poverty? What is the link between health, poverty, and social outcomes? Underscoring the severe health inequalities in modern society, Politzer et al. remind us that "poor people can expect to live shorter lives with more chronic illnesses, and black children are two and a half times more likely than white children to die within the first year of life" (2003, p. 296). These stark statistics not only highlight the individual suffering but also the societal impact of health disparities. Moreover, those living in poverty are more likely to live in a marginalized neighborhood; there is a statistically significant association between neighborhood disadvantage and poor health, which persists even after controlling for personal characteristics (Ross et al., 2001; Wen et al., 2003).

It is widely documented in medical sociology, public health, and social epidemiology studies that economically and socially marginalized groups have worse health outcomes than their more advantaged counterparts, despite relatively high standards of living in some regions of the world. This phenomenon is especially prevalent in the United States, where income inequality is high, meaning that those with better access to financial resources also have better access to healthcare options when they fall ill. The downside is that the opposite is also true—those with limited access to financial resources also have limited access to options to improve their health status. As a result, these populations face poorer health outcomes, placing additional stress on emergency and hospital resources, and ultimately affecting the healthcare system as a whole. It is important to note, however, that the cause of the problem is not just financial. Individuals with limited financial resources often also have limited social resources and connections. This underscores the need for a holistic approach to healthcare that takes into account not only financial but also social factors. As one's financial situation improves, so does their access to other individuals with better resources, better doctors and health resources, and more informed social networks. Think, for instance, of the rollout of cutting-edge medical technologies. Although individuals belonging to historically marginalized groups are likely most in need of transformational medicine, the first recipients of medical innovations are those who can afford the treatments out of pocket and are referred by a medical concierge or private physician. In the late 1990s and early aughts, Link and Phelan (1995) formally named this phenomenon of social factors as the fundamental cause of disease, coinciding with a bevy of studies connecting social class and health.

Notably, Ross et al. (2001), Politzer (2003), Wen et al. (2003), and Kim et al. (2006) found that individuals living in poverty are more likely to live in a marginalized neighborhood or have limited social networks—two factors that are significantly associated with poor health, even after controlling for the effects of personal characteristics. More specifically, individuals inhabiting lower social positions, such as those belonging to historically marginalized social groups, such as racial/ethnic minorities, often face barriers to quality healthcare, including lack of financial access, poor physical access to care, language barriers, and racism. It is important to note that these effects are not just external. Social barriers such as racism or xenophobia also cause physiological effects on the body, as the excess exposure to stress as a result of one's social identity has a clear deleterious physiological effect on the body (Pearlin, 1989; McEwen, 2006). The factors influencing individual health are not singular in nature but multifaceted. While biology and genetics are important precursors to the variation in health outcomes for individuals, social conditions are also fundamentally important as they disproportionately influence individuals' exposures to health risks and protective health factors.

How, then, are these ideas related to where one lives? Although individual characteristics such as race, education, and poverty status are known to have significant effects on health outcomes, where one lives also has a significant effect on health (Macintyre et al., 2006; Klinenberg, 2015; Farmer, 2010). Neighborhood disadvantage is a particularly important predictor of health as it represents the level of social order and control in a neighborhood, which can have an indirect effect on the health of residents. The spatial isolation due to distrust and fear, limited access to healthy food options and green space for outdoor exercise, and the stress-induced lived experience of low-income neighborhoods have long been linked to adverse health outcomes for individuals beyond that of personal disadvantage. This outcome is also attributable to the concept of social capital and levels of exposure to protective and risk factors of disease. Social capital, as defined by the types and depth of social connections that individuals have with one another, provides the opportunity for shared resources and networking that can help or hinder the health of individuals. An individual's level of social capital depends on the size of their networks and the level of trust and reciprocity within these networks. Individuals with high social capital tend to have better access to health resources and social support from friends, family, and neighbors, which could help them cope with a medical crisis, compared to those with similar social characteristics but low social capital.

For Community Health Centers, Access Is the Key

In addition to these social and environmental factors contributing to poor health outcomes, individuals and groups historically relegated to the bottom of the social stratification ladder face barriers to quality healthcare, including financial challenges, transportation issues, language barriers, and dispassionate care, which ultimately impedes their ability to get the medical help they need promptly. Studies suggest that the removal of these barriers, or even the perception of their removal, has marginal or better benefits for historically marginalized populations as compared to similar populations with the barrier to access intact (Saver & Peterfreund, 1993; Casanova & Starfield, 1995; Andrulis, 1998; Lynch et al., 2000). Generally speaking, widening access to quality and comprehensive primary healthcare options, which include a range of services from preventive care to treatment of chronic conditions, helps break poverty-related cycles of disease and death, especially in locations where regular access to doctors is scarce. An early and salient example is offered by Ayanian et al. (1993), who explored the case of breast cancer mortality and insurance status among thousands of female patients in the late 1980s. They found that uninsured women and those with Medicaid insurance generally present later for care than women with private insurance, resulting in the former group having more advanced cases of breast cancer when diagnosed. Moreover, using the New Jersey State Cancer Registry and hospital discharge data, the authors found that the risk of breast cancer death was 40 percent higher for women with Medicaid and 49 percent higher for uninsured women as compared to women with private insurance. They conclude that access to a primary care physician, which is easier to obtain for the privately insured, increases the likelihood of early detection screening so that cancer is caught earlier in the disease cycle. They also suggest that hospitals with a high number of Medicaid and indigent patients may not have the technology for complex diagnoses, potentially resulting in less effective treatment plans.

As this study suggests and hundreds that have followed, individuals in need of care but do not have access to treatment and testing go on living with undiagnosed conditions, pushing through the symptoms until they physically cannot ignore them anymore due to pain or functional failures. When the individual does finally overcome barriers and present for care, their condition has often progressed and requires more acute treatment than initially warranted. Furthermore, this exchange occurs in an emergency room rather than a doctor's office, placing additional more stress and financial burden on the medical system as a whole. It is also important to highlight the issue of insurance. Contemporary debates over the merits and constitutionality of the Affordable Care Act call into question whether all Americans should require health insurance, noting its critical importance to individual health and the stability of healthcare systems. Long-term uninsured individuals are shown to have more unmet health needs than the insured. They are more likely to go without care due to cost and are significantly less likely to have preventative care and diagnostic disease screening. Ayanian et al. (1993) were early recognizers of the need for a multidimensional approach to the improvement of access beyond just insurance, concluding that "Medicaid coverage alone—without efforts to enhance primary care and screening—may be insufficient to improve outcomes" (p. 330). Thirty years later, this conclusion holds and serves as a central concern for the contemporary community health center program.

Access, affordability, and holistic care are at the core of the community health center model, addressing three of the most common barriers to healthcare for historically marginalized communities, groups, and individuals. Consistent with evidence-based data, the program seeks to acknowledge and address the financial barriers to quality healthcare in place for many historically marginalized groups experiencing poverty. A signature objective of the community health center program has been the free or low-cost provision of primary healthcare services for patients. A common mantra in the community health center space is that no patient is turned away due to an inability to pay. Another critical characteristic of the community health center program is the holistic provision of care focusing on all aspects of a patient's health, including physical, dental, mental, and behavioral health services, nutrition education, essential pharmacy services, and social services. The intent is to provide basic access to treatment, advice, support, or referrals to patients for any ailment in a comprehensive and culturally competent manner. The final way that community health centers address the problem of health disparities is through their persistent lobbying of Congress for funds for expansion, offering hope for the future of healthcare equity. Program advocates aspire to make community health centers available in every community where they are needed. With thousands of sites nationwide, the size and geographic reach of the program have grown significantly since its humble beginnings as a demonstration project with two sites.

Community Health Center Program Origins

Johnson's War on Poverty

The formal establishment of community health centers emerged from the War on Poverty legislation, an innovative social policy program. The national fixation on the issue of poverty, coupled with the activism of the 1960s, provided the financial opportunity for community health centers to formalize and receive federal funding under the guise of poverty prevention and management, an effort distinct from the realm of public health efforts. The War on Poverty was a federally funded social experiment to address the "causes and consequences" of poverty in the United States. H. Jack Geiger is widely credited as the founder of the community health center program; an academic and civil rights activist, Geiger was deeply concerned about health inequalities and the social causes of poor health. His model for indigent care in the United States, proposed in 1964, was heavily influenced by Sidney Kark and Guy Steuart's (1962) model of community-oriented primary care in South Africa. Geiger visited Kark and Steuart's site multiple times and was intrigued by the model, which considered everyone in a defined area a patient and consequently received specialized medical care and education to improve the health outcomes for the entire community. Geiger thought this model could greatly benefit medically vulnerable populations in the United States but did not have a clear path to its implementation and funding.

The mid-twentieth century brought a heightened social awareness of poverty, encouraged by a strong culture of activism precipitated by the civil rights and anti-war movements. In 1959, Robert J. Lampman, an economist who would later be considered one of the intellectual architects of the War on Poverty, wrote a report to Congress warning of the slowing decline of poverty rates since the end of the war. Lampman argued that a national poverty rate of more than 20 percent was not only damaging to those living in poverty but also to society as this population is much less likely to be active contributors to the workforce and tax base. The public's preoccupation with economic inequality rose in a period of relative prosperity for the middle class after World War II, illuminating class distinctions. Popular books, such as Galbraith's (1963) *The*

Affluent Society and Harrington's (1962) *The Other America*, also reminded US Americans that the suburban affluence and prosperity enjoyed by many strongly contrasted with the equally real, but often overlooked, poverty and economic hardship of many others, especially those living in rural towns and distressed urban areas. The policy beginnings of the War on Poverty effort can be traced back to the Kennedy administration. Kennedy's brother, Attorney General Robert Kennedy, was sincerely interested in rectifying juvenile delinquency. Through meetings and site visits with nonprofits, police departments, and localized task forces, Robert Kennedy came to believe that the causes of juvenile delinquency were structural, as opposed to behavioral or rooted in personality defects. Robert Kennedy subsequently concluded that community action was the best solution to address the "powerlessness and poverty" that precipitated delinquency in marginalized communities. Heeding his brother's concerns, President Kennedy planned to use the eradication of poverty as the re-election platform for his second term. This goal was never realized as a result of his assassination in 1963.

When Lyndon B. Johnson assumed office following the assassination of President Kennedy, he continued to push for a federal response to the problem of poverty. In his first State of the Union address after President Kennedy's assassination, Lyndon B. Johnson declared a "war on poverty" and emphasized the need for specific social programs, including medical coverage for those experiencing poverty, expanded food assistance, a higher minimum wage, federal aid to education, improvement of public facilities, and mass transit in economically distressed communities. In his defense of the program, President Johnson argued that they offered a "hand up and not a hand out" to constituents but was careful to place ultimate responsibility for the program's success on the public. Further framing this issue as a moralistic and societal duty, in 1963, the *New York Times* devoted an entire article to defining poverty in modern society, concluding, "In any society a premium will be paid to those who are most productive . . . but we need to give people a chance and set a floor of suffering" (Miller, 1963, p. 105).

The rationalization for federal intervention on the issue of poverty was most explicitly stated in Johnson's 1964 Economic Report of the President, in which he argues, "We pay twice for poverty: once in the production lost in wasted human potential, again in the resources diverted to coping with poverty's social by-products. Humanity compels our action, but it is sound economics as well" (Johnson, 1964a, p. 56). In this sense, Johnson recognized both the social justice issue of income inequality and the economic benefits of improving people's economic situations so that they could be positive contributors to American society. Removing the structural forces exacerbating the causes and consequences of poverty gave those belonging to historically marginalized groups a chance to break their perceived cycle of social disadvantage.

In the end, Johnson's bill passed swiftly through Congress with bipartisan support, mainly to memorialize Kennedy. The swift passage of the Economic Opportunity Act of 1964 (EOA) was viewed by many as a way to memorialize President Kennedy, as the bill itself was relatively vague in terms of the types and scope of the programs that would be included under its umbrella. The omission of specific details was a function of the lack of precedent and an effort to be as neutral as possible to maximize bipartisan support (Gillette, 2010). Sardell (1988) argues that "it was a policy initiated within the executive branch, with Congress having a minimal role, little input from interest groups, and little attention from the general public" (p. 50). Moreover, the design of the War on Poverty program was an experimental effort, as there was no blueprint for this type of comprehensive policy for legislators. The EOA did, however, specify the establishment of the Office of Economic Opportunity (OEO) as the federal department that would provide oversight for the programs funded through the bill. A unique characteristic of the effort was that the bill funded individual programs to be implemented and managed by private and nonprofit entities outside of local, state, and federal government. As a result, centralized oversight and management became critical to ensure the prudent use of public funds and the confidence of Congress. The bill itself was relatively vague, but there was intense congressional pressure to allocate funding quickly so that results could be swiftly achieved and reported to the media. This unique sociopolitical environment provided the perfect opportunity for community health centers to formalize under the guise of poverty prevention and management. Situating its mission within the context of a larger action plan to address the severe poverty in the community, community health center program founder H. Jack Geiger and his colleague Count Gibson initially requested $30,000 f a feasibility study in Mound Bayou, Mississippi, and were granted $300,000, ten times the amount requested. Weeks later, the budget was increased to $1.3 million, and the community health center program was formally established as a War on Poverty program (Sardell, 1988).

The OEO

The OEO, a critical component of Johnson's plan to eradicate the causes and consequences of poverty, had a diverse range of missions. The function of this new federal agency was to award funds and provide centralized oversight for the local programs selected to be a part of the War on Poverty initiative. Due to the broad guidelines for the program, the OEO had significant autonomy and flexibility in allocating an initial budget of $65 million. The missions of the OEO-funded programs were varied but had a unifying element of alleviating economic hardship for those living in poverty. With an administration primarily comprised of appointees, nearly any local, grassroots program with federal connections, a concrete plan, and a justifiable reason why federal funds could help alleviate the causes or consequences of poverty had little trouble getting the agency's attention.

The establishment of the OEO was swift, reflecting a sense of urgency and commitment to combat poverty. Sargent Shriver, Kennedy's brother-in-law, was the first OEO Director, and Jack Conway, the former director for industrial unions at the AFL-CIO, was in charge of community action. Sandy Kravitz, a social worker with a PhD in Public Policy, was responsible for day-to-day operations and vetting potential funding recipients. The agency's establishment was rapid and heavily informed by nontraditional federal partners such as grassroots activists, academics, and community-based interest groups, historically ignored in the federal policymaking process. OEO programs also had many passionate and influential advocates who pressured Congress to support their causes.

The first community programs receiving funding under the OEO were Foster Grandparents, Legal Services, and Head Start, all established in late 1965. Recognition of the need for healthcare services came a few months later, as administrators of these early programs noticed that many users had untreated medical conditions, which impacted their abilities to participate in and benefit from the program offerings consistently. For example, over 25 percent of the children participating in the newly established Head Start program had not ever seen a doctor and had untreated severe medical problems, making it difficult for the children to meet the academic and developmental milestones of the program. Despite this troubling finding, Head Start administrators were weary of incorporating direct medical care within their programmatic scope due to budgetary concerns. This hesitation left a critical space for the community health center program to enter the OEO arena and address this need.

The OEO programs were distinct from other federal public policy programs primarily due to the requirements of community action and personal involvement as integral strategies to combat poverty. An essential requirement for OEO programs was the engagement of communities in shaping the agenda of the programs, empowering them to address their unique challenges. This approach also provided job creation, training opportunities, and professional experience for individuals living in impacted communities.

Criticisms of the War on Poverty and the OEO

While the War on Poverty program was generally viewed as a net positive and a fitting way to honor Kennedy, there were initial concerns about its long-term sustainability. Critics from the left argued that the money trickling down from OEO to the local community was insufficient to make a real difference in improving the quality of life for individuals. They believed that the War on Poverty efforts treated the symptoms, not the fundamental causes of poverty. As designed, the War on Poverty programs functioned within existing inequitable structures, focusing on increasing access to resources and promoting positive social behaviors on the individual and community levels instead of disrupting the structures through which social inequities persist. Critics from the left believed that the proper comprehensive solution for resolving the issue of poverty lay not in throwing money at flash-in-the-pan efforts but in the serious consideration and treatment of the structural causes of poverty, notably racism and discrimination in the labor market.

Critics from the right predictably argued that the War on Poverty effort was solely motivated by politics and saw the direct federal-to-local agency link between citizens and the federal government as an over-extension of the federal government's role and diminution of states' rights. Administratively decentralized with the exception of funding and occasional reporting requirements, OEO programs initially called on educational institutions, nonprofits, and community groups to implement federally funded programs with minimal federal guidance and the circumvention of state and local governments, ruffling some congressional feathers. A particularly iniquitous objection arose from Southern legislators at the state level as they perceived the requirement of community involvement, coupled with the lack of state control, as dangerously providing economic support and political power to Black communities and organizations and thereby weakening the white power base in their jurisdictions (Sardell, 1988). At this time, the dissolution of the Jim Crow Era and the rise of proposed racial equality through the Civil Rights Act left many southern legislators feeling disempowered on the federal landscape. More locally, the mere mention of community involvement in program administration was met with scorn by conservative legislators who perceived a threat to their power as individuals and groups traditionally left out of the policy and political process were now legislatively empowered and not only could now vote thanks to the Voting Rights Act but also had a direct say about how and when federal funds were allocated for most OEO programs. To alleviate this concern and build goodwill, the compromise was to give the states some veto power in selecting the OEO programs established in their jurisdictions. However, OEO had the power to override any specific objections based on the social and economic needs of the area.

The most significant objection to the War on Poverty and OEO efforts was the fear that the program would erode the personal responsibility of those living in poverty. Many in Congress and the general public believed that personal responsibility and the rejection of mainstream values were at the core of America's poverty problem. In this widely held view, individuals and communities in poverty were there due to their poor personal choices and values. The recent passage of the Civil Rights Act was seen as a step toward a post-racial society. With the denial of fundamental rights based on social characteristics now illegal, everyone was presumed to be on an equal playing field toward economic success.

However, instead of making serious and sensible life choices, those living in poverty, especially Black and Brown people, were perceived as stubbornly and foolishly creating an underdeveloped subsociety, allowing these populations to stay in poverty and survive on the government's benevolence and exist despite lacking the necessary resources to participate effectively in the larger society (Gillette, 2010). In this worldview, spending more public funds on a never-ending cycle of poor choices was nonsensical. The Watts riots of 1965 also reinforced this negative perception of Black and Brown communities being incorrigible and undeserving of the compassion and benevolence doled out by the War on Poverty effort. Daniel Patrick Moynihan, a prominent Republican legislator and the Assistant Secretary of Labor at the time, formally captured this perspective in a widely circulated government report, "The Negro Family: The Case for National Action," also known as the Moynihan Report (Office of Policy Planning and Research United States Department of Labor, 1965). In it, Moynihan argues that the disproportionally high poverty rates in the Black community compared to the rest of American society were a result of their cultural predisposition to social dysfunction rooted in their lack of family structure. This culture-based deviance resulted in a perpetual cycle of generational poverty and anti-social behavior that cannot be resolved by government intervention. Quoting E. Franklin Frasier, a Black sociologist, and expert on the Black family, Moynihan asserts in the report:

> As the result of family disorganization a large proportion of Negro children and youth have not undergone the socialization which only the family can provide. The disorganized families have failed to provide for their emotional needs and have not provided the discipline and habits which are necessary for personality development. Because the disorganized family has failed in its function as a socializing agency, it has handicapped the children in their relations to the institutions in the community. Moreover, family disorganization has been partially responsible for a large amount of juvenile delinquency and adult crime among Negroes. Since the widespread family disorganization among Negroes has resulted from the failure of the father to play the role in family life required by American society, the mitigation of

> this problem must await those changes in the Negro and American society which will enable the Negro father to play the role required of him. (Office of Policy Planning and Research United States Department of Labor, 1965, Chapter V., para. 9)

From this perspective, the War on Poverty Effort was futile at best because people are the problem, not systems, and the fundamental problem is the failure of the Black family. Without the discipline and socialization required to contribute to American society successfully, increased access to resources would be of minimal benefit to this subpopulation. Therefore, the premise of the War on Poverty effort, in the view of critics from the right, is a misguided and costly approach to addressing the issue of poverty. President Johnson famously argued that the War on Poverty program represented a "hand up and not a hand out" to citizens in need in his public defense of the program. He promised that if given a chance, the social access provided by the initiative would be transformational for those experiencing poverty. To back up this claim, once the program and the OEO were authorized, the agency quickly and strategically released funding to several programs at the beginning of the funding cycle so that results could be achieved quickly and reported to Congress. However, measuring the success of a program with so many different potential outcomes and macro-level conflating factors such as the economy, wages and interest rates, demographic shifts, and policy changes became a significant challenge for the OEO. Soon after the programs were up and running, Johnson became increasingly distracted by the Vietnam War and murmurs that the War on Poverty initiative was "wasteful, mismanaged, counterproductive, and a cause of social and racial turmoil" became a common complaint in legislative discussions (Gillette, 2010, p. xiii). In response, the OEO instituted an internal Office of Inspection that reported directly to the OEO Director and provided audits of the programs to minimize fraudulent use of funds. Although there was some internal dissent within the agency, the perception of strong operational oversight and control coupled with measurable outcomes was critical for the program's ongoing support.

Although the OEO lasted only a decade before being incorporated within other federal departments, virtually all its components have survived for over fifty years under ten different administrations. Since the implementation of the War on Poverty initiative, the US poverty rate dropped from 22.4 percent in 1956 to 12.6 percent in 1970. It continued to hover in the low teens through the beginning of the twenty-first century. This trend is undoubtedly perceived as improvement, but not the policy juggernaut that was promised or expected. Because it was eventually disbanded, it is difficult to isolate the effects of the OEO program efforts from the general economic trends that contributed to the decline in the poverty rate over time. The complicating effects of racism, systemic inequalities, and vacillating public opinion on public welfare spending over the decades to follow have also obscured the more extensive program's effects. However, despite the debate around its tangible benefits, the program has cemented its place in history as one of the few net-positive attempts by the federal government to address the ills of poverty in the United States on a national scale, leaving an enduring legacy that continues to shape policy and public discourse.

The First Centers

In the summer of 1964, H. Jack Geiger served as a field coordinator for the Medical Committee for Human Rights, a grassroots nonprofit providing health services to civil rights protestors and residents in the rural community of Mound Bayou, Mississippi. By the Fall, Geiger had independently initiated a small community clinic to better serve the staggering medical needs of those in that area. At the time, health outcomes for Blacks in Mississippi were comparable to an underdeveloped country. Due to extreme poverty and racism, many residents of Mound Bayou had never seen a medical doctor; public and private hospitals in the area were segregated, and the availability of Black physicians was limited. With limited options for quality care, many in the Black community eventually stopped trying to access traditional medical care at formal facilities. Instead, they relied on home remedies, which had a disastrous effect on health disparities in the region (Bailey & Danziger, 2013).

Despite the challenges, Geiger and Gibson's perseverance was unwavering. In need of funding and a sustainable medical model for the community clinic, Geiger drew upon the historical success of the dispensary model and contacted a former colleague, Count Gibson, Chair of Preventative and Community Medicine Chair at Tufts Medical School, to explore a residency model. He proposed that Tufts provide partial funding and oversight for the newly launched health center in return for residency and training spots at the center for medical students, giving Tufts, a Northeastern school in an urban setting, a competitive advantage in offering medical students the experience of training and practicing at a southern, rural, community-based center. Gibson agreed to the partnership. For the remaining funding, Geiger and Gibson pitched the surgeon general of the Department of Health Education and Welfare (HEW) for federal support to augment the funds provided by Tufts. Their timing perfectly coincided with the establishment of the OEO, to which HEW immediately referred them. Although the duo initially requested $30,000 for a feasibility study, OEO granted $300,000, ten times the amount requested. With the additional funds, Geiger and Gibson expanded the Mount Bayou plan to include an urban health center in South Boston near Tufts in the Columbia Point neighborhood.

The medical landscape in Boston was different but just as dire as Mound Bayou. In Columbia Point, there was extreme poverty and medical need in the shadow of the elite medical and academic institutions of the city. Charity care from major healthcare institutions was fragmented and unwelcoming to Black patients, especially those living in poverty. Like Mound Bayou, there was also a dearth of private physicians outside hospital outpatient systems, limiting options to those in need. After adding Columbia Point to the project, the increased scope of the project, coupled with the OEO's need to award funds in the first year expeditiously, resulted in another significant increase of Geiger and Gibson's budget from $300,000 to $1.3 million within a matter of weeks. The OEO's support was instrumental in the program's expansion, demonstrating this organization's influence in the indigent healthcare landscape at the time.

Despite the seemingly unlimited supply of funds, not everyone was enthusiastic about the federal support of community health centers. R. Sargent Shriver, a member of the Kennedy family and prominent politician credited with founding the Peace Corps and Head Start programs, was skeptical about the community health center program, especially after noting the objections presented by the AMA. To satisfy his doubts, Shriver enlisted the then-leader of the Head Start program to weigh in on the feasibility of Geiger and Gibson's project. The administrator, who was also a physician, supported the need for the health center and thus saved the program from being cut before fund allocation. The community health center program funds were approved and distributed in June of 1965, merely six months after the initial conception. Within the next eighteen months, the OEO enrolled and funded health centers in eight US cities.

Massachusetts quickly emerged as the epicenter of the community health center program due to Columbia Point's high visibility, direct affiliation with connection to an elite university, and congressional support. Formally a home visit program operating from a large housing project, Tufts repurposed the program into a healthcare facility. Academics, legislators, and the new media heralded the project as an immediate success. The opening of the community health center resolved the problem of limited access to quality healthcare for thousands of impoverished residents of the housing project. A year after the center's opening, Senator Ted Kennedy became one of the program's most vocal and constant supporters. With Kennedy's help, the community health center program was authorized two years later, in 1967, with a $51 million earmark; thirty-three new centers were funded in the following year. This program had a profound impact, providing quality healthcare to those who had previously been underserved.

In 1968, the US Public Health Service (PHS), the federal public health agency distinct from the OEO, began funding health centers through their agency in an attempt to leverage the success of the OEO's community health center program. Traditionally, the PHS "avoided programs in areas in which organized medicine was sensitive: programs that might influence the delivery system of medical care in the United States, and programs that involved the federal government directly in financing continuing integrated care to significant numbers of Americans" (Sardell, 1988, p. 69). Strong medical lobbies and political powers required public health agencies to stand down often and bend to the general will of the AMA and traditional medical-based interest groups. As a result, the PHS health center effort reframed the discussion about the public benefit of health centers away from poverty eradication and toward coordinating national healthcare resources across both the public and private sectors. This narrow approach resulted in a relatively small funding base for PHS health centers compared to the OEO program centers. The PHS health center program remained relatively small until it was merged with the OEO program during the Ford Administration.

Five years after the initial disbursement of program grants from the OEO, there were 100 centers receiving funding. In the decades to follow, various bipartisan legislations skyrocketed the rapid expansion of the community health center program. In the 1970s, the Emergency Health Personnel Act authorized the NHSC program, increasing medical professionals' availability in medically underserved communities. The "medically underserved" designation, primarily reserved for communities with low physician-to-population ratios and then later expanded to include the consideration of the percentage of people living in poverty, infant mortality rates, and the percentage of elderly residents in an area, would become a powerful indicator in determining which communities were most in need of federal support for healthcare infrastructure and resources. The NHSC program offered doctors willing to relocate to and practice for at least five years in these areas financial remission for their medical school loans in exchange for their service. This initiative contributed to an influx of physicians in underserved areas, promoting the growth of community health centers in new, underserved areas. There is, however, mixed evidence of whether physicians stayed in these areas after the initial reimbursement period. Nonetheless, many nonprofits, including churches, community centers, and hospitals, took advantage of these new eligibility requirements and applied for new access point grants. In 1975, Congress formally authorized community health centers as 'community and migrant health centers," reflecting the centers' commitment to various at-risk populations, including unhoused and precariously housed individuals, migrant workers, and those with limited English proficiency. The "medically underserved area" designation was also renamed "health professional shortage area" and became a required designation for most social programs receiving federal assistance, most notably the community health center and Head Start programs. After ten years as a demonstration project, the community health center program also received permanent authorization in the 1970s.

The community health center program received additional boons in the late 1980s and early 1990s as Congress authorized the current FQHC designation for the centers, reflecting their new eligibility for cost-based reimbursement from Medicaid, malpractice coverage under the Federal Claims Tort Act, and deeply discounted drug costs through the newly established 340B prescription pricing program. These were substantial financial wins for centers as they could now be made "whole," receiving a full-cost reimbursement for every Medicaid patient seen in their centers, low costs on prescription pricing, and subsidized malpractice coverage for medical providers. The favorable reimbursement rates and program savings also helped to offset the cost of treating uninsured and underinsured patients and the cost of providing enabling services such as case management, outreach, health education, nutrition, and translation services. Legislation favoring the growth of community health center programs continued through the 1990s as federal reimbursements for Medicaid patients soared with the establishment of a prospective payment system for health centers and the creation of the State Children's Health Insurance Program.

The new millennium ushered in some of the most impactful legislation, spurring the exponential growth of the community health center program and its permanence as a safety net provider. The Obama/McCain presidential race marked the first time that both Democratic and Republican candidates publicly endorsed the community health center program's expansion in a Presidential race. Soon after his victory, President Obama made the most significant investment in the community health center program in history through the American Recovery and Reinvestment Act of 2009 and the Affordable Care Act of 2010. Through these two legislations, the community health center budget grew by over $13 billion over five years as it was established as a critical partner in the broader healthcare system. The criticality of the program's role was reinforced in the response to Covid-19 a decade later as Congress approved $7.6 billion for community health centers as part of President Biden's American Rescue Plan of 2021. In FY 2023, the federal allocation to the community health center program budget was just over $5.5 billion.

Although the growth of the community health center program has been a net positive over the past six decades, AMA lobbyists and advocates have kept a watchful eye on the structure and role of the program. With one of the most organized and powerful lobbyists, the AMA has been persistent in its requirement that the community health center program not infringe upon the deeply established prominence of traditional healthcare providers. In the early years, the fight was about redundancy and scope. The AMA saw the community health center model of integrating primary care and public health as redundant, wasteful, and disregarding longstanding boundaries between public and private medicine. Later, fairness and competition became concerns as the increasingly generous funding and reimbursement environment specific to community health centers were deemed unfair by some private physicians and pharmacist advocates, especially given that community health centers receive both grant monies and Medicaid reimbursement rates to make them whole. Community health center advocates and defenders have successfully ameliorated these concerns and others in the legislative space, carving out and maintaining a unique role for community health centers as a safety-net option serving the entire healthcare structure and remaining generally nonthreatening to the traditional healthcare model.

Initial Program Regulations

Like the legislation that established the OEO, early guidance from the federal government on the health center program regulations was vague. Initial legislation specified that affiliated centers would provide a wide range of ambulatory services, be accessible, involve the community, collaborate with other agencies, and prioritize Medicaid and Medicare funding. Otherwise, there were few specific guidelines. During the program's first four years, medical schools and hospitals with outpatient centers received most of the funding. However, the federal requirement of meaningful community involvement soon became a complicating barrier for these larger institutions that were not accustomed to external forces shaping their work. Local health departments also struggled to comply with the requirement of comprehensive care, as the centers were more accustomed to functioning as urgent care centers and were weary of encroaching on the turf of private practitioners.

Capitalizing on this tension, community-based activists fought to separate community health centers from academic institutions and hospitals, emphasizing the need for community need to shape and drive the program, as opposed to institutional missions and goals. This emphasis on community involvement is crucial, as it makes the audience feel engaged and connected to the program's mission. Community health center administrators and advocates ultimately won this battle due to the OEO's documented commitment to the community-centered model. Advocates subsequently pushed for the requirement that centers belonging to the program have a majority-user advisory board or governing board where users comprised at least one-third of the committee. These new requirements prompted most hospitals, medical schools, and local health departments to opt out of the program. Unfortunately, once separated from larger institutions, many health centers experienced strained relationships with their previous institutions. Community health center-affiliated physicians found it challenging to obtain admitting privileges and specialty care for uninsured patients (Bailey & Danziger, 2013). Professionalization of the community health center staff also became a concern once larger institutions abandoned the program, leaving less experienced administrative staff to manage operations. The OEO subsequently offered governance training to health center administrators and community board members to help alleviate this problem.

Conclusion

Federal efforts to address public health have historically fallen under the jurisdiction of the US PHS. For nearly a century, this department's efforts have included disease prevention and treatment, the management of outbreaks and epidemics, and educating the public about avoiding disease and death through positive behaviors. However, the community health center program model was not established as a public health initiative. Instead of focusing on the spread of disease, its goals were to build communities and eradicate poverty by helping ensure vulnerable populations had the proper resources, education, and job opportunities to address health problems blocking the path to social success and achievement. The activism of the 1960s, coupled with a national fixation on the issue of poverty, provided the financial opportunity for the community health center program to formalize through the OEO under the guise of poverty prevention and management—an effort distinct from public health efforts. War on Poverty programs, of which the community health center program was one, further differentiated themselves from traditional federal social programs primarily through the required components of community action and personal involvement to eradicate poverty. For these programs, community engagement, leadership, and the creation of local jobs and training opportunities were common strategies to support local economies and help raise people and neighborhoods out of poverty.

Simultaneously, various federal agencies and associated interest groups have been increasingly concerned about persistent community-level health disparities between mainstream whites and historically marginalized social groups. However, it is essential to note that hostile social forces such as racism, xenophobia, homophobia, and classism have traditionally been at the root of shaping the federal approach to indigent health policy. This has resulted in the magnification of individual and cultural behaviors as the primary cause of disparities, and precipitated policy solutions that were careful not to threaten the comfort of mainstream populations or traditional models of healthcare. Consequently, the responsibility of providing healthcare to medically vulnerable populations, such as those that have been historically marginalized or without sufficient access to healthcare, has been positioned as a special case or issue falling outside the standard approach to care provision. This approach reinforces the othering of communities, placing their voices and needs at the margins of the health policy process. Therefore, it is logical that community health center advocates and administrators have gradually modified the perception of the program goals over time, from providing healthcare to empower communities and eradicate poverty to providing healthcare to fortify the nation's healthcare system and reduce the aggregate cost of uncompensated care.

This action is consistent with good leadership; organizations dependent on federal subsidies must be responsive to the expectations of public opinion and presidential agendas. The original goal of community empowerment, however, remains a beacon of hope for the future impact of the program. Evolving the program's social narrative to maintain alignment with the external environment allows it to maintain its core mission while managing stories about its role and purpose to foreground its benefits to the greater public is reasonable for organizational success. However, this approach also has drawbacks as the needs of historically marginalized populations are further de-centered, resulting in the persistence of relatively poor health outcomes for these populations. As long as historically marginalized populations continue to be placed on the back burner of the American policy agenda, the root causes of social inequities impacting these groups will never be fully addressed. In the case of the community health center program, it should be celebrated for having administratively withstood the budgetary and political challenges to its existence. At the same time, there is a cautionary lesson in its story as those living in neighborhoods around community health centers no longer have them as a primary source of community empowerment as the centers prioritize their work to bolster the larger healthcare infrastructure over the provision of care, education, and jobs with a primary purpose of fighting poverty.

Note

1 For continuity, the term FQHC is interchangeable with community health center in this book unless explicitly referring to the federal designation label.

2

Dressed to the Nines

Social Image and Policy Administration

The US policymaking and federal budget process emulate an intricate dance, reflecting a myriad of interests, checks and balances, and the broader political and economic environment. The complexity of the process is shaped by the active involvement of various levels of government, multiple stakeholders, and fierce advocates. These entities interact at all levels of the policy process, identifying issues and concerns that impact the public and proposing solutions for action and funding. The influence of public opinion, interest groups, and expert analyses, as communicated through the media, further complicates the process. A crucial part of the policymaking process is the proposed budget, a similarly convoluted process heavily influenced by legislative and public opinion about the importance of a proposed program. Each year, the President submits a budget request to Congress, outlining the administration's priorities for federal spending and revenue. This proposal sets the tone for the upcoming fiscal year but does not dictate final expenditures. Congress then reviews the request, debates spending levels across various government programs, and works through appropriations committees to allocate funds. This process often involves trade-offs and compromises among different sectors of government, political ideologies, and regional interests, and is deeply influenced by fiscal constraints such as the federal debt and deficit.

Once Congress agrees on a budget, the appropriations process begins, where specific funding levels are determined for each program or department. Throughout this process, various committees hold hearings, gather data, and negotiate to secure funding for priority initiatives. The need to balance competing interests, address the national debt, and comply with legal constraints makes the budget process a high-stakes, time-sensitive endeavor. This chapter delves into some of the intricacies of the US policymaking process that have impacted and influenced the development of the community health center program over the last sixty years. It begins with an overview of federal public policy development, focusing on two well-established models in public administration: the issue network and the iron triangle. It then touches on the federal budget process and the complexities that require a delicate administrative dance of maintaining legislative relevance and favor while upholding programmatic ideals. The chapter ends with the strategic partnerships between public agencies and the media to build public and legislative awareness of programs and to shape opinions, ultimately impacting legislative perception and decision-making.

Public Policy Development: A Complex Landscape

The policy process in the United States is profoundly complex and often paradoxical. While several theoretical models seek to clarify the steps and flow of the process, it is best described as an organized anarchy consisting of ever-changing lists of priorities, negotiations, emotions, and legislative successes and failures at all levels of government (Cohen et al., 1972). By design, it is open to any constituent who wishes to participate, and any of those constituents is empowered to add a new issue to the list as the democratic process is intended to make space for the ideas, preferences, and needs of everyone, regardless of social or political position. However, despite this openness to new participants and ideas, the relative power of those participants causes some voices to be heard louder than others, leaving the resolution and implementation of those ideas in a seemingly perpetual legislative to-do pile until more powerful social forces demand change. As a result, the US policy process is highly resistant to change despite relatively high levels of transparency and representation built into the system.

The US democratic system, intentionally tempered by a system of constitutional checks and balances, is intended to resist the whims of individual actors and actions in the policy process and promote the mobilization of public opinion before sweeping changes. Fueled by anti-statism, the Founders vehemently opposed a powerful, centralized government, having experienced and subsequently divested from the crushing pressures of living under a strong, centralized government. They subsequently designed our democratic system to maintain consistency and the status quo while protecting individual liberties—a strong foundation for weathering economic and social factors that could disrupt the stability of governmental functionality. The Founder's vision is especially evident in the Executive branch, as Norton Long (1949), a seminal voice in the field of public administration, reminds us that "the Presidency, like Congress, responds to and translates the pressures that play upon it. The real mandate contained in an Executive order varies with the political strength of the group demand embodied in it, and in the context of other group demands" (Long, 1949, p. 257). In this sense, the nation's founding documents, most notably the Declaration of Independence, Constitution, and Bill of Rights, manifest the gentlemen's hopes for a new way of governing and freedom from the English monarchy that still shapes our government today. The Founders' built-in restrictions on the scope of government also left room for the eventual establishment and growth of the nonprofit sector to manage and support social functions perceived to be outside the government's direct purview but still necessary to ensure the social functioning of society, such as caring for the common good. Despite its exclusionary imperfection and short-sightedness in terms of diverse race, gender, and socioeconomic categories, the Founder's vision and the documents establishing the rules of government and law supporting that vision have endured intensive scrutiny and existed as living documents for centuries.

While several established models describe the broad strokes of policymaking in the United States, the framework of issue networks versus iron triangles offers the most basic structure for understanding how policy decisions about the community health center program have progressed in the second half of the twentieth century. Issue networks are a dynamic network of individuals, interest groups, and organizations that come together to discuss and influence policy issues in modern government processes. Interactions can be transitory, meaning individuals enter and exit the process based on their level of interest and engagement. More importantly, in this policymaking model, individual participants' motivations are often noncomprehensive and are rooted in emotion instead of expertise or specialization. For instance, someone may participate in a policy discussion about abortion only because they care about the right to life, but not the social, economic, medical, or emotional consequences the pregnant person may endure as a result of proposed abortion restrictions. In an issue network, that individual is still welcome to participate even though they do not fully understand the nuances of their position or have a recommended solution for any unexpected consequences of their preferred policy action. As a result, issue networks are characterized by patterns of cooperation and conflict where a wide variety of inconsistent actors preclude sustained cooperation without resolving conflict. Gridlock is naturally also common.

Despite its challenges, the issue network model of policymaking offers several benefits. With the widespread use of social media and easy participation, issue network participants are diverse in their priorities, experiences, and perspectives. This diversity allows elected officials to hear directly from constituents and vice versa, fostering a more inclusive approach to addressing social issues. In his classical article *The Science of "Muddling Through,"* Lindblom (1959) discusses the value of having multiple perspectives at the table when making policy decisions. Although his framework centers on the policy decision-making process of the individual public administrator, his theory of successive limited comparisons and its "good" policy results are applicable more broadly in the contemporary policy environment. On addressing complex social problems, he asserts, "On many critical values or objectives, citizens disagree, congressmen disagree, and public administrators disagree. . . . Administrators cannot escape these conflicts by ascertaining the majority's preference, for preferences have not been registered on most issues; indeed, there often are no preferences in the absence of public discussion sufficient to bring an issue to the attention of the electorate" (Lindblom, 1959, p. 81). The result of this complexity is a problem of value prioritization and balancing the relative power of the participants in the decision process. Essential to this framework is the recognition that each participant has organizational and person-specific missions, values, and goals that stay intact during the policy process. Within this context, sound public policy satisfic balancing a maximal combination of the needs of all participants, where each party can live with the resolution without feeling their core identity has been sacrificed or lost. This concept of muddling through is particularly relevant to issue networks, where the diverse and transient nature of participants necessitates a similar approach to policy decision-making.

This contemporary approach to policymaking is distinct and evolved from a traditional model characterized by a much more insular and exclusionary method of interaction on the federal level. Up through the mid-twentieth century, the US policy process was characterized by a rigid, closed subsystem called the Iron Triangle. The Iron Triangle primarily consisted of a handful of interest group advocates, legislators, and key agency administrators cooperatively negotiating policy options and preferences. Iron triangle participants were well-known to one another and worked together on an ongoing basis. Each invited participant represented the wishes of their factions but also shared a common set of policy goals and objectives (Berry, 2010). There were clear rules of engagement—temporary consensus was common to keep programs and initiatives alive as participants conceded to one another with promises of more favorable consideration in the next round of negotiations. Political favors, logrolling, and the threat of mission creep characterize this approach to policymaking, as each participant commonly gives up parts of their needs for the sake of the group. Despite these sacrifices, coming to a consensus in a closed system is much quicker than an issue network, but the quality of the outcomes can be low.

The national school lunch program, a Second World War food program promulgated by a closed collaboration between the US Department of Agriculture, select legislators primarily interested in reinforcing the farming industry, and a small group of school advocates, is a well-known example of the iron triangle model. For decades, these three groups have enjoyed ongoing interactions of low visibility to influence school lunch policy, where all actors shared interests in promoting the program and a direct or tangible concern about how the policy was shaped. They also all had an area of specialization and or technical knowledge relevant to the policy options and the long-range effects of policy decisions on diverse issues such as student health, the farming economy, and educational budgets. Healthcare policy was also traditionally determined by an iron triangle of influential stakeholders. This closed group of interested parties commonly included the AMA and insurance lobbyists, Department of Health program heads, and various interest groups representing voices of public concern. As time passed, the policy process became more public in general, and healthcare policy negotiations evolved into an issue network with many unconnected hubs of participants across sectors, levels of government, and communities.

The Federal Budget Process

Another essential concept for understanding the context behind the persistent success of the community health center program across Presidential administrations is the recognition that public budgets are both administrative and political documents. Federal budgets are not just about allocating dollars and cents; they are about priorities—statements about who and what is important and how these priorities are ordered within the context of competing interests and priorities (Rubin, 2000). These characteristics make budgets inherently political, as interested parties view them as a means to garner attention and secure priority consideration. Rubin explains, "Budgets reflect choices about what government will and will not do. They reflect a general public consensus about what kinds of services governments should provide and what citizens are entitled to as members of society" (Rubin, 2000, p. 1). The impact of these choices on marginalized populations, such as those served by the community health center program, is significant and underscores the need for a thorough understanding of the budget process.

As the needs and demographics of the United States became increasingly diverse, consensus about the role of government became increasingly fleeting on both a legislative and public sentiment level. Today, the federal budget is an incredibly sensitive issue that requires careful consideration and planning on the part of legislative decision-makers advocates, and the public alike. It is the main instrument through which the government allocates resources to various programs and initiatives, and the decision-making process involves a complex interplay of political, economic, and social factors. It is a complex and intricate process involving a wide range of stakeholders and actors and is the backbone of the government's finances. Because it determines how taxpayer money is allocated across different sectors and programs, it is essential that publicly subsidized agency leaders and their advocates understand and engage with this process to ensure that organizational narratives are accurately managed, marginalized voices are heard, and public priorities are accurately reflected in the final budget.

The budget process in the United States begins with the President's proposal, setting the tone for future budget discussions. The President's proposal is a significant document that outlines the administration's fiscal priorities and policy initiatives. It is then up to Congress to review and refine this proposal, taking into account the needs and desires of constituents, lobbyists, and public opinion. This is where the structure of interests for various issues becomes involved, each party advocating for the programs and sectors they believe should receive funding. The US budgetary process is inherently contentious for this reason, as these different stakeholders have varying priorities and objectives that they want to see reflected in the allocation of funds.

Additionally, economic changes and other external factors can significantly impact the budget process, making the budget a constantly evolving and fluid document. Like the policy development process, the demands and voices associated with competing interests and priorities can become louder and more abrasive during the budget process, depending on the available funds and external sociopolitical factors such as recession or war. Active participants seek to ensure that taxpayer money is used in a way that benefits the most people, reflects the nation's values, and supports those in need, but only to the extent that the public deems those populations cannot help themselves. This is where education and advocacy can be particularly beneficial for administrators. However, in doing so, policymakers, administrators, and advocates must approach the budget process with sensitivity and awareness of its potential impact on various groups and sectors of society.

Effective budgeting requires a deep understanding of the needs and priorities of various stakeholders, as well as a willingness to engage in constructive dialogue and compromise to achieve a balanced and fair outcome. Ultimately, the sensitivity of federal budgets underscores the importance of transparency, accountability, and democratic representation in policymaking. As explained in the next chapter, the federal budget process becomes even more fraught when considering the role of government in healthcare and the provision of healthcare to historically marginalized populations that may be unable to pay for quality services. Not since Johnson's War on Poverty had the federal government taken a clear stance on the prioritization of the causes and consequences of poverty, including indigent healthcare. Not only is it challenging to determine which combinations of agencies should control and be responsible for these expenses, but cross-sector coordination is also of interest and open to debate. Questions such as how much the individual should be responsible for their own healthcare, regardless of their ability to pay, become a critical assumption to be determined as part of the budget process. Or, similarly, how much uncompensated care should private hospitals be expected to absorb before the federal or local government provides relief? Or, specific to the case of the community health center program, what level of care is the federal government willing to subsidize to meet a minimum level of care that is enough to leave an individual stable enough to be a productive member of society? These budget questions are more about values, norms, and social priorities than financial allocations and permeate community health center budget conversations.

The Critical Role of the Public Agency Administrator

Formal definitions of public administration as a field of academic study and profession are voluminous and varied, but generally coalesce around the idea that public administrators' work involves managing and coordinating resources to implement public policy fairly and efficiently. Public policy initiatives, such as providing quality healthcare to those with little to no access, traditionally seek to distribute government resources to eligible constituents. While elected officials create policies to best serve the public good, public administrators are responsible for implementing them fairly and in a manner that treats all constituents as equals. When applied with a social justice lens, as the field of public administration requires, implementing policy initiatives can help restore equal access to public resources when the government has historically failed a particular social group in that area. Administrative responsibility calls for the contemporary public administrator to pursue social equity as both a means and an end. The tension between equality and equity aligns with the oft-employed and classical Friedrich versus Finer debate, a theoretical discussion that compares two different approaches to public administration. Friedrich (1940) calls for public administrators to rely on their technical expertise and let professional responsibility guide their administrative decisions. Finer (1941), on the other hand, emphasizes the administrator's subordination to the elected official and accountability to the will of the public. Both approaches are evident in modern administration, but the former is associated with a commitment to equity and social justice, even if public sentiment wavers in its support. Nonetheless, regardless of the intent, equitable outcomes are less attainable than outcomes rooted in efficiency or economy, as political and budgetary pressures require prioritizing procedural fairness over the promotion of justice (Gadson & Park, 2024).

Public administrators operate within all three branches of government at all levels of our democratic system. They are also at the helm of civil society, serving their communities in the nonprofit and broader public sector. Across the world, public administrators lead agencies and activities that operationalize government actions and laws in a way that directly impacts individuals and communities within a prescribed area or jurisdiction. To do so effectively, they must respond to a complex mix of professional, personal, social, and legal obligations, adding to the challenge of leadership in the public sector. Moreover, public organizations are like fishbowls. Their missions, budgets, and activities are vulnerable to environmental influences and subject to intense scrutiny by external stakeholders, such as legislators and the general public. This is partially because taxpayers, funders, and donors have expectations of transparency and effectiveness when public dollars are involved. In this sense, the boundary between the internal organization and the external environment is paper-thin, and only the most effective administrators can successfully balance the pressures of both environments. Social and economic trends can also intensely impact the demand and scope of public services, resulting in the need to adjust internal operations and staffing to respond to the changing expectations of external stakeholders. In their roles, public agency administrators manage this decision process and implement any plans resulting from those decisions. They are also expected to maximize efficiency, effectiveness, and fairness outcomes when planning the distribution of public resources while simultaneously keeping operational costs, such as overhead and marketing, as low as possible.

While public administrators are undoubtedly leaders in the realm of operational management, the extent to which they should take a leadership role in the policy process is an ongoing debate in the field. In the purist's sense, public agency administrators are primarily responsible for implementing public policy in their role. To ensure the fair distribution of resources, public policy administration must be free from political interference. Therefore, policy development falls wholly within the elected official's scope of responsibility. Woodrow Wilson famously coined the term politics-administration dichotomy to describe this truism in the late 1800s as he formalized the profession as an academic discipline and insisted that "the field of administration is a field of business. It is removed from the hurry and strife of politics" (Wilson, 1887, p. 209). Once legislators make policy decisions on behalf of the public, they are handed down to the public agency administrator, who then brings the policy to life by applying their technical expertise with public systems and organizational leadership. In this model, the administrator's primary job is to ensure that the policy initiative is implemented objectively and efficiently on behalf of the government or funders. This means that public agency administrators are not responsible for creating policy, but rather for executing it in a way that is fair and just. This dichotomy separates politics, which is concerned with policy formulation, from administration, which is concerned with policy implementation to ensure the fair and just application of public policy, especially at the federal level. For Wilson, the administration process is subordinate to and separate from the political process. By keeping politics and administration separate, the government and the public can ensure that administrators act in the public's best interests, rather than being influenced by their political connections.

While professional ethics and longstanding laws such as the Hatch Act of 1939 point to the inappropriateness of administrators responsible for public funds and programs directly participating in electoral politics or attempting to influence the outcome of a democratic process, it is nearly impossible to separate politics from administration completely. Instead, effective administrators must partner with legislators and elected officials to best meet the needs of constituents. Wilson's politics-administration dichotomy provides an essential framework for ensuring accountability and transparency in government, but does not fully capture the interconnectedness and complexities of today's policy environment. As this book suggests, the community health center administrator carries a substantial and heavy load of ensuring that their patients and communities are heard and represented in the policy process, and that both legislators and the public are appropriately informed as they formulate their opinions about the program, especially during times of budget reauthorization.

Strategic Plans for Administrative Success

As early as the 1940s, public administrator scholars acknowledged that the role of the administrator was not just technical but also normative, political, and creative. Many leaders of public agencies enjoy a level of discretion as they are trusted to efficiently and effectively implement public policy utilizing the necessary creativity to translate government directives into meaningful initiatives that meet the specific needs of their communities. Choosing how to prioritize and balance the various objectives often associated with public policy is also a political process. F. M. Marx (1949) writes, "There is always the inescapable question of the political effectiveness of public administration, perhaps better called its social efficiency. The only meaningful way in which the social efficiency of the administrative system can be judged is by testing its operations for conformity with the fundamental valuations that inhere in the political order" (Marx, 1949, p. 1127). It is from this foundational framework that a critical link between the public administrator and the elected official can be established and cultivated, despite Wilson's call for separation. For a public agency to be deemed successful, it must be perceived as progressing the values and goals of the entity or individual assessing its progress. Managing this relationship is a political act without which a publicly subsidized agency cannot survive, as it loses value and falls out of alignment with expected norms and priorities.

Focusing on the contemporary nonprofit landscape, Salamon (2015) underscores the strategic role of nonprofit administrators in ensuring the survivability of their organizations. He points out that the nonprofit sector as a whole is losing its distinctiveness and power, facing unprecedented challenges due to the current economic climate, government funding cuts, and increased competition. To overcome these factors, Salamon urges nonprofit administrators to adopt a more strategic and purposeful approach to their operational plans, diversify their funding sources, and collaborate with other organizational leaders. Notably, he emphasizes the importance of helping policymakers recognize the vital role of nonprofits in society and provide appropriate support and resources to help the sector thrive.

Strategic planning plays a crucial role in managing the image of nonprofits, particularly in politically and socially challenging environments. It is a "deliberative, disciplined approach to producing fundamental decisions and actions that shape and guide what an organization (or other entity) is, what it does, and why" (Bryson, 2018, p. 8). Given the critical gaps nonprofits fill in society, their survival cannot be left to chance. Environmental scans serve as an essential tool for informing strategic plans. They help organizational leaders systematically identify, deeply reflect, and creatively respond to the opportunities and threats presented by the external environment. From these scans, administrators can develop the necessary operational tactics to identify and refine organizational strengths and weaknesses.

The purposeful nature of the strategic planning process to ensure agency survivability is particularly relevant to the community health center program's quest for survival across multiple Presidential administrations. External communication has been instrumental in their success, with lobbying and media coverage being their most effective tactics for maintaining relevance and securing funding. In the conclusion of his 2015 text on the challenges facing the nonprofit sector and what they must do to reverse these trends, Salamon asserts,

> The public must be introduced to the broader realities of the current nonprofit operations, to the impressive resilience that the sector exhibits, and to the special qualities that make nonprofit organizations worth protecting. Also

> required is a better public defense of the sector's longstanding partnership with government and a clarification of the crucial role nonprofits play as key links working collaboratively with government and the business sector to solve public problems. (Salamon, 2015, p. 111)

Salamon's advice closely aligns with the actions of community health center administrators and advocates over the past six decades. They have consistently strived to be part of the political discourse on healthcare and poverty support and engaged the public using the media to be champions of their cause. They also forged crucial legislative and media relationships for coalition-building and to generally raise awareness about the community health center program and how its goals align with those in power in Congress and at the White House.

Media Construction of Moral Narrative

Norton Long (1949) has long established that the "lifeblood of administration is power," underscoring that public agency survival depends on the ability to influence power structures in society and government (p. 257). Wielding influence requires the management of stakeholder relationships, paying particular attention to the most relevant mix of the agency's structure of interests, which comprises various stakeholders such as elected officials, media, interest groups, and public sentiment with varying levels of interest in the organization's causes. Therefore, using media as a communication tool is a smart administrative strategy for shaping organizational image and messaging perceptions. In addition to amplifying the voices of diverse stakeholders, the media performs a critical function in the policy process by educating the public and raising awareness of policy debates during times of significant change. Astute public administrators and advocates of public programs understand that the structure of interest requires ongoing cultivation and influence to keep an organization viable and relevant in the policy landscape and the sociopolitical environment.

The policy process is increasingly transparent and participative, but also requires significant advocacy and the active political involvement of constituents to create, implement, and support large-scale policy change effectively. The media serves as a critical mechanism for connecting the government and the public during the policy process, as it excels at educating the public and raising awareness of policy debates in times of significant change. As previously discussed, day-to-day policy concerns within government are navigated by relatively closed policy subgroups primarily composed of legislators, interest groups, lobbyists, and advocates. Because our democratic system promotes the mobilization of public opinion before making any sweeping changes, significant policy changes occur only when routine issues rise to the level of public consciousness. Traditional media has historically been the mechanism through which this occurs. Policy administrators and their organizational advocates can leverage the media's professionalism and reach to bend policy narratives toward positive public sentiment for their causes.

Legislators, who propose bills, authorize spending, and approve resolutions, members of the public who hold legislators accountable, and the media, which define the context and translate the messages between these two groups, are profoundly intertwined in the policy process. Because day-to-day policy negotiations tend to be a relatively closed process negotiated within policy subsystems, it is not until an issue reaches the larger public consciousness due to a shift in the macro-environment that the public becomes an active participant in the policy process (Weible, 2012). The media has the critical role of sifting through numerous issues and concerns simultaneously under discussion at the legislative level to determine which issues require public attention and the appropriate tone and scope for communicating them. All of this takes place within a social and political system in which the public's attention span is short. Striking when the proverbial iron is hot is a crucial strategy for the media, advocates, and legislators alike (Croft & Wagner, 2018).

The media's role in determining what policy issues should be brought to the public's attention and the subsequent framing of those issues for consumption have been long established (McCombs & Shaw, 1972). A civic-oriented approach to journalism is particularly beneficial in the public sector, as it engages in the education of dominant social groups about the lived experience of marginalized populations to "foster a sense of communal responsibility" that can set the tone for reproducing or challenging current policy and institutionalized inequality (Hodgetts et al., 2008, p. 62). The media is part of an organization's structure of interests and interpretive community: a group of individuals who share similar interpretations and understandings of the world, creating shared meaning through personalizing news stories and issues to inform culturally competent public opinion (Berkowitz & TerKeurst, 1999).

The ongoing partnership between the media and issue advocates has raised awareness about critical social issues and promoted positive change for centuries. By working together, media outlets and advocates can amplify important messages and ensure they reach a wider audience. Nonprofit agencies, in particular, establish an "arena of action through which individuals can take the initiative not simply to promote their own well-being but also to advance the well-being of others" (Salamon, 2015, p. 16). This partnership is not only effective, but also necessary when the media needs to sound the alarm in cases of injustice and amplify the need for public solidarity. In this way, the partnership between the public sector and the media has proven effective in bringing attention to various issues, from environmental concerns to social justice causes. This partnership must continue to thrive to create a more informed policy environment.

In the case of the community health center program, administrators and supporters have historically partnered with the media to shape specific narratives about community health centers and their patients in a way best suited to the given political and social environment (Sardell, 1988). In these instances, constructing moral narratives that can influence future policy action is a successful approach to building public and legislative support. Malone et al. (2000) explain that the

> journalistic telling of any story, no matter how "objectively" presented . . . encourage[s] the reader of the story toward a particular understanding of the "facts" of the story, and make predictions regarding the direction that the issues may take in the future. To the degree that this normalizing discourse characterizes what "many people" think, feel, or believe, it also transmits commonly held social values and provides suggestions for how conflicts among values might be resolved. (Malone et al., 2000, p. 728)

Co-creation of a subjective reality based on the objective reporting of the facts is a powerful tool for shaping public opinion about public spending and governments' responsibility to marginalized populations. As later described in the *New York Times* analysis in Chapter 3, the media is an indispensable partner and tool for community health center administrators and advocates to humanize the intensity of the need for centers to remain funded and operational. This humanization is a key aspect of the media's role in shaping public perception and policy outcomes.

Conclusion

The public policy landscape in the United States is a dynamic and intricate system. Since George Washington's first inaugural address in 1789, each newly elected president has set a unique tone for their time in the White House. Some, like President Obama, have championed the ideals of unity in their leadership. Others, like President Clinton, have focused on creating a leaner and more efficient government. However, these presidential agendas are not unilaterally determined. Our democratic election process ensures that they are aligned with and rooted in the wishes and priorities of the public, a testament to the wisdom of the Founders' enduring vision. Public sentiment and opinion, at the core of the elected official's sense of responsibility, play a pivotal role in shaping the policy landscape. The Founders designed a system that necessitates elected officials to incorporate the desires and needs of the voting public in the policy process. This empowers citizens to influence the country's policy direction through their votes and voices.

Due to the fishbowl effect and public scrutiny, it is prudent for public agency administrators to dedicate resources to ensuring that the public and the elected officials who represent them have a positive and accurate image of public programs. The media is a critical tool for sharing this narrative and social image. In his theory of communicative action, Habermas (1984) asserts that effective communication is not just beneficial for the two-way transmission of information. It can also facilitate mutual understanding and agreement between parties without the need for domination and coercion.

This framework aligns with the mutually beneficial relationship between community health center administrators and the media. In addition to public education about happenings and events, community health center administrators have leveraged news stories to connect with stakeholders and build partnerships and understanding. They communicate the program's mission, values, and benefits through storytelling about their centers, patients, and communities. These narratives create connection, familiarity, and access for stakeholders, who can become policy advocates both formally and informally. This is one example of what Habermas identifies as the lifeworld in civil society: the everyday social interactions shaped by shared norms and values across institutionalized sphere of modern society., c

While the contemporary policy process is messy and complicated, it ultimately leads to more inclusive policies that better meet the needs of the public. Both the Iron Triangle and the issue network model have their pros and cons. The Iron Triangle is predictable and clean, but change is often elusive. On the other hand, the issue network process is messy and complicated, but that is also its beauty. It is unpredictable as interested parties move in and out of the process; however, it leads to a more inclusive policy that better meets the needs of the public than a few hand-selected parties often far removed from the populations for whom they are making decisions. In today's political environment, the inconvenience and messiness of the issue network are necessary for ensuring a more open and just policy process.

As discussed in later chapters of this book, the transparency and openness of federal policy supporting the community health center program have led to numerous administrative successes for the program. Community health center administrators, advocates, and congressional supporters use effective communication techniques to manage expectations about the program, the populations they serve, and the program's need for adequate funding to achieve organizational goals, benefit communities, and support the broader healthcare infrastructure.

3

Dancing the Two-Step

Shifting Social Narratives for Political Survival

Barring the recent exception of the Covid-19 pandemic, curative health initiatives have traditionally commanded more legislative and public attention than preventative public health efforts, receiving 97 percent of the more than $3.5 trillion spent annually on health in the United States (Faberman et al., 2020). In routine times, Congress and the AMA mostly relegate preventative healthcare to the realm of private providers and hospitals for most Americans, with community health centers positioned as a federally subsidized safety net in historically disfranchised communities (Raudenbush, 2020). With its unique positionality in the healthcare funding landscape, the community health center program represents a unique mix of curative and preventive healthcare models. Its foundation is rooted in prevention; however, the ability of community health centers to identify, diagnose, and holistically treat primary care ills to reduce uncompensated care costs to the broader healthcare system has been an important key to its legislative survival. In this sense, the program's purpose extends beyond the realm of health to that of economic security for the nation, placing it within the responsibility and scope of federal responsibility and attention.

Over the past sixty years, support for the community health center program has been consistently subject to the changing landscape of federal politics. No president since Johnson has made poverty eradication a national priority, although healthcare for indigent populations has received some specialized attention in several administrations. In the United States, traditional media have historically been the primary mechanism through which routine public sector concerns rise to the level of public consciousness. Most of the time, the policy process is insular—routine lobbying, persuasion, and education of elected officials outside committee meetings happen in offices and informal engagements behind closed doors outside the purview of the public. However, when significant policy changes are under congressional consideration, the policy process becomes more transparent and subject to public criticism. In the case of the community health center program, a robust policy network comprising activists, interest groups, and key legislators has sustained the program's funding over the years. During reauthorization, community health center program administrators and their advocates carefully craft their messages to highlight the relevance and importance of their programs and position the organizations as aligned with the current sociopolitical environment. As delineated by the presidential administration, there are ten significant periods of authorization associated with the community health center program from 1965 to 2024, as shown in Table 3.1. During these periods, legislation impacting the budget, scope, or rules for community health centers became law. For each period, the Presidential administration, years covered by that administration, the associated dates used to search and categorize the *New York Times* articles, and the name and brief description of the legislation's impact on community health centers are provided in the table.

This chapter systematically explores the evidence of the narrative shift in the depiction of the community health center program in response to the social, economic, and political forces present during these legislative periods. The analysis is based on a comprehensive content analysis of legislative language in bills contributing to the reauthorization and funding of the community health center program, as well as media articles using a sociological lens. Each chronologically ordered subsection is punctuated by milestones in the budget or program reauthorization of the community health program within the larger context of the presidential agenda of the era, as determined by the newly elected President's first joint address to Congress. The implications for the community health center program, as well as examples of the program's depiction in the media and legislative documents during these periods, are also detailed in each section. On the legislative side, Presidential agenda-setting not only serves as a proxy for public sentiment about the role of government but also provides context for the legislative environment within which administrators

Table 3.1 Major Community Health Center Program Authorization Periods by Presidential Administration

Presidential Administration	Years	Community Health Center Impact	Associated Legislation	*New York Times* Coverage Dates
Lyndon B. Johnson	1963–9	Initial funding as a ten-year demonstration program	Economic Opportunity Act of 1964 (Public Law 88-452)	22 November 1963 to 19 January 1969
Gerald R. Ford	1974–7	First authorized as a permanent program	Health Services and Centers Amendments (Public Law 94-63)	9 August 1974 to 19 January 1977
Jimmy Carter	1977–81	Major expansion and doubling of funding	Health Services and Centers Amendments of 1978 (Public Law 95-626)	20 January 1977 to 19 January 1981
Ronald Reagan	1981–9	First real threat of elimination defeated and expansion secured	Temporary Emergency Food Assistance Act of 1983 (Public Law 98-8); Community and Migrant Health Centers Amendments of 1988 (Public Law 100-386)	20 January 1981 to 19 January 1989
George Bush	1989–93	Establishment of Federally Qualified Health Center (FQHC) status and corresponding benefits	Ryan White Comprehensive AIDS Resources Emergency Act of 1990 (Public Law 101-381); Federally Supported Health Centers Assistance Act of 1992 (Public Law 102-501)	20 January 1989 to 19 January 1993
Bill Clinton	1993–2001	Consolidation of community, migrant, and homeless health center programs	Health Centers Consolidation Act of 1996 (Public Law 104-299)	20 January 1993 to 19 January 2001
George W. Bush	2001–9	Implementation of "no one turned away" policy and sliding fee system	Health Care Safety Net Amendments of 2002 (Public Law 107-251); Health Care Safety Net Act of 2008 (Public Law 110-355)	20 January 2001 to 19 January 2009

Table 3.1 (Continued)

Presidential Administration	**Years**	**Community Health Center Impact**	**Associated Legislation**	***New York Times* Coverage Dates**
Barack Obama	2009–17	Passage of Affordable Care Act and $11 billion in additional funding	Patient Protection and Affordable Care Act (Public Law 111-148); Health Care and Education Reconciliation Act of 2010 (Public Law 111-152)	20 January 2009 to 19 January 2017
Donald J. Trump	2017–21	Subject of Senate hearing and Inclusion as primary partner in federal Covid-19 response	Access to Care: Health Centers and Providers in Underserved Communities (Senate Hearing 116-276); Continuing Appropriations Act, 2020, and Health Extenders Act of 2019 (Public Law 116-59); Further Continuing Appropriations Act, 2020, and Further Health Extenders Act of 2019 (Public Law 116-69); Consolidated Appropriations Act, 2021 (Public Law 116-260)	20 January 2017 to 19 January 2021
Joseph R. Biden	2021–4	Infusion of billions of dollars in restitution following the pandemic	American Rescue Plan Act of 2021 (Public Law 117-2)	20 January 2021 to 14 February 2024

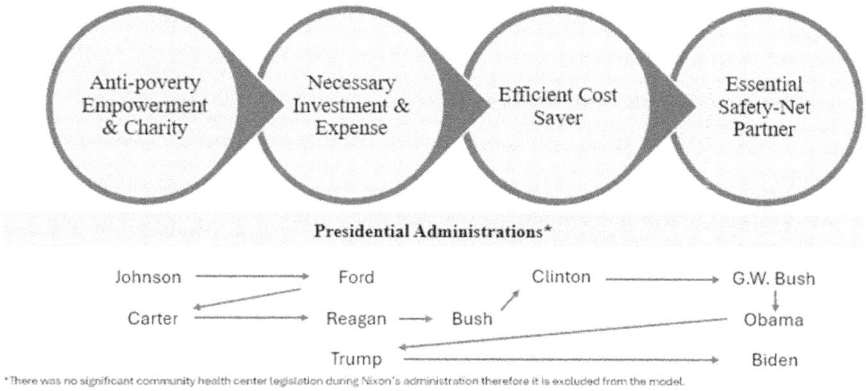

Figure 3.1: Community Health Center Program Narrative Shifts Across Presidential Administrations: 1965 to Present. *Source*: Author

and advocates must contend. From a public opinion perspective, the media serves as an indispensable two-way communication partner between program administrators and the public. As the evidence suggests, while the mission of the community health center remains consistent, the perception of its social function and public good has shifted from its inception as a means for poverty eradication to its current role as a safety net for the larger healthcare system. The four roles cross administration periods and are explained in Figure 3.1 and below:

- **Anti-Poverty Empowerment and Charity:** As part of a broader poverty-eradication effort on the federal level, community health centers positioned themselves as the main source of primary healthcare services, health education, nutrition, and jobs for their local communities. As a group, their programmatic purpose was to strengthen neighborhoods and eradicate poverty by ensuring people are healthy and stable enough to effectively participate in society and better resist the causes and consequences of poverty.
- **Necessary Investment and Expense:** Positioning themselves as a PHS, community health centers highlight their ability to provide culturally competent care and fill a crucial gap in the broader healthcare system that no other organization aspired to fill in this stage due to stigma and low insurance reimbursement rates, such as care for LGBTQIA+ or immigrant

populations. Within this framing, poverty becomes an indicator of need rather than a target of elimination, and the focus is more on the individual rather than on communities.

- **Efficient Cost Saver**: Community health centers position themselves as fulfilling a critical cost-saving role for the larger health center system by managing the demand of Medicaid and uninsured populations, thereby diverting these groups from emergency rooms and uncompensated care. This strategic approach significantly reduces healthcare costs for the entire system, underscoring the program's economic value.
- **Essential Safety-Net Partner**: Community health centers have fortified their position as unique and integral members of the broader larger healthcare landscape, reducing costs and expanding primary care access for individuals with various barriers to care.

Generally, these narratives are marked by forward progression across presidential administrations. However, some transitions retreat to an earlier narrative, such as between the Ford and Carter administrations and the Obama and Trump administrations. In these instances, the program narratives returned to a previously proven legislative and public support strategy as a well-worn administrative tactic.

The following sections trace this narrative shift in community health center perception across the ten reauthorization periods (note that there was no significant community health center legislation during Nixon's abbreviated Presidency, thus it has been omitted from this analysis). Each subsection describes the sociopolitical context of the time, as defined by the Presidential administration and the initiatives outlined in the first State of the Union address. Following a brief discussion of the significant legislation impacting community health centers during the period, a thematic review of more than 230 corresponding *New York Times* articles highlighting the work of community health centers or their patients further contextualizes the public and legislative perceptions of community health centers. Details of the methodological procedure, as well as the justification for the use of the *New York Times* as the primary media source, are provided in the Methodological Appendix.

Johnson: Poverty in the Cross Hairs

A commitment to rid society of its social ills predominantely shaped President Lyndon B. Johnson's time in office. Following President John F. Kennedy's footsteps, President Johnson's extensive list of domestic goals included advancements in civil rights, tax cuts for the poor and middle class, poverty reduction, healthcare reform, education, and the general improvement of the quality of life for all Americans. He used his administration to connect widespread social opportunity to the economic stability and growth of the nation, suggesting that helping those struggling the most rises the boat for everyone. His first State of the Union set the stage for his War on Poverty effort, emphasizing the federal government's responsibility to combat and eliminate poverty, promote equal opportunities, and work toward eliminating racial discrimination. In his speech, he asserts, "Unfortunately, many Americans live on the outskirts of hope—some because of their poverty, and some because of their color, and all too many because of both. Our task is to help replace their despair with opportunity" (Johnson, 1964b, section III para. 1). He then outlines economic development, job creation, tax reforms, and expanded social security and healthcare plans to reach this aspiration. Highlighting the importance of unity and faith and the degrading nature of hate following the assassination of President John F. Kennedy, President Johnson ends his speech with a call for bipartisan efforts to achieve these social objectives for the benefit of all citizens.

President Johnson's commitment to eradicating poverty was not a mere vision but a tangible reality that materialized during his administration. His tenure was characterized by significant social spending, with Congress authorizing the establishment of the first federally supported community health centers through the EOA of 1964. These centers, positioned as beacons of hope for poverty-stricken neighborhoods, provided avenues for education, training, and healthcare through local nonprofit organizations. The legislation, which later became Public Law 88-452, included provisions for financial assistance, free healthcare, student work-study programs, and community action programs aimed at addressing poverty. The Act also established the

OEO and outlined various guidelines and regulations for effective program implementation. Legislative support for the initiative generally aligned with Johnson's administrative goals, and the public fervently supported any effort that posthumously honored John F. Kennedy. Massachusetts Senator Ted Kennedy, known for his liberal politics, was elected to Congress in 1962 and began his rise in prominence during Lyndon B. Johnson's administration. He would become a lifelong champion and vocal advocate for the health center program. Community health centers were also closely affiliated with universities during this time, further safeguarding them from legislative threats.

Community health centers, a key component of President Johnson's War on Poverty, were first introduced to the nation by advocates like Kennedy. Their role was to underscore the need for quality healthcare in economically depressed neighborhoods and demonstrate how these centers could help alleviate the causes and consequences of poor health at the community level. By providing accessible and affordable healthcare, these centers played a crucial role in improving the health and well-being of the residents in vulnerable neighborhoods, thereby contributing significantly to the overall poverty reduction strategy of President Johnson's administration. Once passed, the federal guidance gave health centers and their associated partner organizations the freedom and flexibility to use the abundance of funds to best meet the specific needs of their target populations and communities. This adaptability ensured that the health center program could effectively respond to the evolving health needs of the community.

During the initial period of the War on Poverty, *The New York Times*' coverage of community health centers contextualized the work of indigent healthcare as charitable and predominantly benefiting populations distinct from mainstream society. The public saw individuals from historically marginalized communities as different and less refined, rendering them incapable of fully participating in society without assistance. For example, an early piece from February of 1966 compels readers to consider disease symptoms amongst those experiencing poverty as abnormal and not to be overlooked as routine. In a piece explaining the need for a new health center in Brooklyn, the author states, "A lot of people think kids are supposed to be round-shouldered, feet are supposed to itch, you're supposed to be depressed every night and wake up with a cough every morning" (Tolchin, 1966, p. 13). The author insists that the public must pay attention and be charitable to historically marginalized communities, rather than turning a blind eye or turning away in disgust. The articles from this period illustrate this common tactic for connecting the general public with marginalized populations for the purpose of charity and the resolution of social problems, mirroring President Johnson's approach in Washington.

Moreover, media articles highlighted the need for charitable donations to enable health centers to pull patients into care and research these distinct pockets of society to better understand the people who live there. As a social reward for those who did give, the newspaper also frequently published short articles to display the names of generous benefactors prominently, such as one highlighting the "$1-million gift from Mr. and Mrs. Louis E. Russo of New York for the establishment of a Community Health Center in the Bronx" ("Einstein Medical College," 1968, p. 72). Stories about medical innovation were also frequent, as "old methods of treating poor children no longer meet the needs of the underprivileged child of the central city" ("Medical Center Opens in Bronx," 1968, p. 26), or the nearly full-page description of the "automated intensive-care unit that completely takes over the chore of monitoring and recording a critically ill patient's pulse, temperature, respiration, and blood pressure" (Bird, 1967, p. 36). The media presented these innovations and the support of community health centers as necessary investments in the fight to improve the quality and scope of indigent healthcare in order to charitably care for those in need.

Ford: Press the Brakes and Shift Gears

Republican Gerald R. Ford's presidential agenda marked a significant departure from President Johnson's intense focus on social policy and poverty eradication. His first State of the Union speech, delivered in 1975, less than six months after the Watergate scandal and his ascension to the presidency, set the tone for his more objective and pragmatic approach to the nation's issues and his proposed solutions. His primary goals, including a tax cut to stimulate the economy and energy conservation measures to reduce reliance on foreign oil, were a clear departure from Johnson's policies. Like Johnson, Ford stressed the need for bipartisan cooperation in his first address, but he was also unequivocal about halting the creation of new programs and reducing spending on social programs to address the current economic challenges and secure the nation's future. Emphasizing this position of restraint, he stated,

> Part of our trouble is that we have been self-indulgent. For decades, we have been voting for ever-increasing levels of Government benefits, and now the bill has come due. We have been adding so many new programs that the size and the growth of the Federal budget has taken on a life of its own . . . when these programs are enacted, there is no dollar amount set. No one knows what they will cost. All we know is that whatever they cost last year, they will cost more next year. (Ford, 1975, para. 24–25)

This statement speaks to the growing public critique of the free-flowing nature of funds associated with the War on Poverty programs, which began a decade prior. Through his administration, President Ford hoped to reinstate the country's global and domestic strength position, calling for a new direction toward solving unemployment, increasing individual and household incomes, and controlling government spending—a contrast to Johnson's priorities.

Luckily, the community health center program continued to have strong legislative support during Ford's administration, mostly attributable to its alliance with powerful higher education institutions in the Northeast, such as Tufts University, and a decade of demonstrable proof of its effectiveness. These assets paved the way toward permanent authorization through the amended Public Service Act of 1975, which updated and expanded various public health programs under the federal government's purview. The amendments to the Act included grants for comprehensive PHSs, family planning, and mental health programs with specific requirements for compliance, cost-reporting, and status performance obligations to provide better oversight and ensure quality assurance of programs. These structured requirements precipitated the first evidence of how the program's perception shifted when administrators, advocates, and the media discussed the community health center program purpose. Moving from an anti-poverty program under the OEO to a public health initiative under the PHS shaped how it was viewed and depicted. This shift in perception, from a poverty-focused program to a public health initiative, shed light on the evolution of healthcare policies and the changing priorities in the healthcare landscape.

Newspaper articles about community health centers reflected this change in depiction. One salient example was the coverage of a center using an innovative telemedicine approach using bi-directional cable television to connect an off-site pediatrician, pediatric orthopedic specialist, and a psychiatrist with patients in need to expand access to care for vulnerable populations. Decades before the rise of telemedicine as we know it, prominent physicians had volunteered their time from a hospital office across town for the better part of two years to provide real-time consultations to nurse practitioners with patients in the room at an urban medical center. The reporter explains, "The installation—one of a handful of such two-way medical hook-ups in the country and the only one designed to serve an outreach pediatric clinic in an inner-city community—is coming down chiefly because its application for federal funding has not been renewed" (Brown, 1975, p. 61). The article goes on to explain that the research derived from the setup has been valuable. Still, the service design does not align with the requirements of the new community health center program. The article quotes an administrator who finds "video consultations by specialists to be inadequate healthcare and has insisted that the clinic have doctors on the premises full time" (Brown, 1975, p. 61). The result is losing a clinic location as program eligibility shifts under the new federal guidelines. Several other articles from this same timeframe covered the search for new funding sources for innovative programs across the county that fall outside the lines of traditional healthcare models.

Before their permanent authorization under the PHS, the community health center program represented scores of demonstration programs employing innovative solutions to address barriers to healthcare in marginalized communities. These innovative solutions, which ranged from telemedicine approaches to the search for new funding sources, demonstrated the program's adaptability and commitment to its mission. In this sense, providing healthcare through the community health center model is a means to eradicate poverty's causes and consequences. Shifting healthcare provision to be the goal, rather than one of many tactics to achieve the goal of poverty eradication on the community level, precipitates a subtle but significant change in the perception of the program. It also signals the federal government's reduced emphasis on poverty eradication through healthcare at the local level.

Carter: Compassion and Realism

Known for his compassion and unwavering commitment to human rights during his tenure as President, Jimmy Carter took office in 1978 during high national inflation and an impending recession. His first Congressional speech emphasized the need for unity and partnership between legislators and the people to address the nation's persistent challenges around energy, job creation, tax reform, and welfare. President Carter and his administration prioritized civil service reform and reducing government bureaucracy, which he saw as integral to the effective partnership between elected officials and the electorate. In his approach, he rejects the idea of the government as the primary arbitrator arbitrator of social justice, instead recognizing it as a critical but single component of a more extensive social system balancing the competing needs of a diverse constituency. In his address, he states,

> We really need to realize that there is a limit to the role and the function of government. Government cannot solve our problems, it can't set our goals, it cannot define our vision. Government cannot eliminate poverty or provide a bountiful economy or reduce inflation or save our cities or cure illiteracy or provide energy. And government cannot mandate goodness. Only a true partnership between government and the people can ever hope to reach these goals. (Carter, 1978, para. 24)

President Carter promoted a culture of an open and honest federal government committed to transparency and accountability to best address the nation's domestic and international challenges. His departure from conventional norms regarding the government's role in addressing social issues did not deter his administration's expansion of the community health center program. Like his predecessors, Carter recognized the program's value as a proven method for delivering tangible social services to underserved communities. However, what set him apart was his personal identification with the severity of inadequate healthcare provider coverage in rural areas, a connection that stemmed from his childhood experiences. This personal empathy and understanding were evident in his commitment to America's rural landscape, manifested in his expansion of the national park system and the additional protection of 103 million acres of land.

Consequently, the Rural Health Initiative was enacted in 1977, effectively doubling community health center funding over a four-year period. This significant increase in funding, a testament to President Carter's commitment to healthcare, was more than just a financial boost. It was a beacon of hope for previously overlooked communities in rural parts of the nation promising improved access to quality healthcare. A year later, Public Law 95-626, enacted in November 1978, further amended the PHS Act to extend financial assistance programs for health and preventive health services to improve government-subsidized health service delivery, particularly for community health centers. It begins the formalization and professionalization of centers by providing grants and contracts to support technical assistance for the planning and operation of centers and implements regulations for primary care centers with centralized oversight mechanisms. The Act also speaks to disparities in access to quality healthcare and supports innovative approaches for primary health services delivery to underserved populations. Overall, the Act aims to enhance health services, prevent diseases, and promote overall well-being in communities across the United States, painting a promising picture for healthcare in America.

The theme of underserved populations being unique and needing support was carried through the *New York Times* coverage of the era. However, the coverage highlights the negative aspects of the differences between the mainstream population and those in need of health, drawing readers' attention to the direness of disparities in healthcare access instead of speaking to the common humanity of people. One article covering the challenges of a senior health coordinator responsible for pulling patients into care began with the descriptively shocking othering of vulnerable populations, stating, "They wait for her, the swollen mothers, the discards with rheumy eyes or a whisky list . . . the wobbly minded ones, emptied out of the state's mental institutions, who haven't yet crawled into a corner to rot" (Sheehy, 1979, p. C2). The author asserts that community health centers served those no other medical institution desired to help, emphasizing their essential social role. The article also emphasizes that she came from the community she served and described her as a "rare sort of leader to emerge from the community of the damned. Out of their solitary hells she leads those whom the rest of society has condemned to invisibility" (Sheehy, 1979, p. C2).

Other articles similarly pointed to the social undesirability of the behaviors and choices of historically marginalized populations, such as the reporting that "half of all black children born today are born out of wedlock," or that "well-run" community clinics have financial difficulties because patients don't pay their bills and seek to game the system and get services for free ("Health-Clinic Strikers," 1978, p. B2). A physician in another article stated, "Let's face it . . . a physician has to make a living, and how many are going to invest the $60,000 or so it costs to open a practice in an innercity neighborhood. He's going to lose money" (Sullivan, 1978, p. NJ1). These narratives, while biased, painted a picture of the importance of community health centers doing the dirty work and thus dutifully proving their worth in federal grants and support.

Reagan: Budget Cuts and Block Grants

From the outset of his presidency, Ronald Reagan committed to economic progress, competition, government accountability, and reducing the size of the federal government to reduce its involvement in constituents' everyday lives. His first address emphasized efforts to reduce government spending, cut taxes, deregulate industries, combat waste and fraud, and restore military strength. He also addressed the need to reform entitlement programs, control government spending, and strengthen federalism, stating that "government will not continue to subsidize individuals or particular business interests where real need cannot be demonstrated . . . we call for an end in duplication of federal programs and reform of those which are not cost-effective" (Reagan,

1982, para. 15). As a result of this mandate, states would gain control of federal grant programs, giving them more autonomy in the administration and distribution of resources to their communities.

Reagan's administration promoted the establishment of urban enterprise zones to create economic incentives and promote economically healthy competition for the funding of social efforts. The federal government proposed returning various social programs and associated funding controls to state and local governments to improve accountability and decision-making. This proposal specifically targeted numerous longstanding social programs from the War of Poverty efforts established decades before and receiving ongoing federal funds, including the community health center program. As part of this aggressive plan, President Reagan proposed splitting the program into voluntary state-controlled block grants with a 25 percent reduction in funds. This recommendation met considerable opposition from legislative supporters of health center interest groups. It became the primary focus of advocacy efforts by NACHC, the national advocacy group representing community health center interests at the federal level. Despite the significant challenges, the community health centers, with the support of NACHC, demonstrated remarkable resilience and determination in maintaining their funding levels and protecting the interests of the communities they serve.

New York Times articles from this timeframe highlighted the impact of cutting services and dramatized the need to support historically marginalized communities with programmatic and fiscal support. One article covering a legislative tour of health centers pointed out that "one client at the Dr. Martin Luther King Jr. Health Center in the Bronx, the largest community health center in the country, looked quizzically at the visitors trooping through the center and asked, 'The President isn't going to close this one down, too, is he?'" (Rule, 1981, p. 27). This quote illustrates the trickle-down effect of the impending bill on individual members of local communities, a powerful commentary on the reach of the cuts. Newspaper coverage also set the context for understanding the dire effects of the cuts. It included quotes from patients not being able to afford alternatives and "waiting until they pass out" before seeking care in a crowded emergency room. It also had information about centers' rejection of uninsured new and newly pregnant patients in favor of patients with insurance, the instance of out-of-pocket visit fees skyrocketing seven to twenty-five times higher than usual, and the unwillingness of traditional facilities to take on the patients that community health centers would have to turn away in the case of budget cuts. The Rule (1981) article also warned that doctors serving marginalized populations were finding their work less than satisfying due to the "pressures the Government places on [them] with cuts and productivity standards" (p. B3).

From an educational perspective, the *New York Times* articles reviewed from this period also reminded readers that indigent care costs do not disappear when the federal government imposes budget cuts. People without proper access to preventative healthcare will continue to fall ill, and the costs of their care will shift from the federal social safety net to state and local governments. There will also be more acute hospital-based uncompensated care, rising costs for the entire healthcare system. In another piece, a healthcare commissioner for a particularly marginalized locality noted, "There are going to be horrendous cuts at all levels of county government as we try to absorb the impact of President Reagan's reductions" (Melvin, 1981, p. WC13). The paper reports the ratio of doctors to patients in her county as 1 to 4,600, compared with 1 to 323 nationally. Moreover, many of the traditional doctors who did practice in the county chose not to accept Medicaid insurance for fear that "taking poor patients will give them a poor image" (p. WC13). The impact of Reagan's policies on the healthcare system was significant and concerning, leading to a shift in the burden of care and a decrease in the availability of healthcare services for the most vulnerable populations.

Ultimately, community health center advocates achieved a significant victory in saving the program from block grants. Congress reauthorized the program in 1988, and Reagan could not find a congressional sponsor for his cost-cutting model. Despite this victory, the program still experienced a reduction in its appropriation from $368 million in 1981 to $321 million in 1982. The cuts disproportionately impacted the individual centers as the administration required them to submit need-based calculations to justify funding allocations. The legislative actions resulted in the phase-out of 187 centers, with others receiving severely reduced funding.

Bush Sr: Getting Down to Business

President George H.W. Bush campaigned on a promise to return America to its traditional values of faith, family, and freedom. During his time in the White House, he faced several challenges to the nation's position as a global power, including ending the Cold War, initiating Operation Desert Storm, and overthrowing Manuel Noreiga's dangerous regime in Panama. In his inaugural State of the Union address, President Bush acknowledged the world's significant changes and challenges and emphasized the need for the United States to promote peace and freedom. For domestic affairs, he committed to an ambitious investment in America's future through education, environmental protection, and economic growth. Instead of scoring historically marginalized populations, he acknowledged the importance of taking care of those with financial struggles so that "every one of us enjoys the same opportunities to live, to work, and to contribute to society . . . live in dignity . . . [and] no American is forgotten" (Bush, 1990, para. 18). Education and healthcare spending became national priorities in Bush's administration, as evidenced by his commitment to Head Start, another War on Poverty program, and the discontinuation of outrageously high healthcare spending through service reduction and containment for programs with federal oversight. Like the Presidents before him, he called for domestic and international unity and cooperation to achieve these goals.

President Bush's administration was a pivotal time for the community health center program, a period during which the program's resilience and adaptability shone through. The business component of center operations was redesigned and formalized, with specific and detailed legislation governing the expected scope and actions of centers receiving federal funds. This marked a significant shift from the health centers' grassroots beginnings. Congress, guided by community health center advocates and allies, spent considerable time determining the optimal configuration for minimizing risks of indigent care to the federal government while ensuring effective operations remain appropriately funded. Efficiency, effectiveness, and accountability were paramount, and in the end, the administration challenged community health center administrators to demonstrate their business acumen and evidence of organizational effectiveness.

New York Times pieces from this period detail the challenges of malpractice insurance for providers who see uninsured or underinsured patients, especially for specialties such as obstetrics or pediatrics. Beyond cost, appointment availability is scarce, as few doctors reserve space for these risky patients. As one reporter explains, "Insurers see poor people, especially pregnant women, as being bad risks. Such people, they say, often avoid contact with doctors and hospitals until their medical problems become acute" (Pear, 1991, A18). Solutions proposed in the article are that centers pool their money to negotiate premiums as a collaborative effort or appeal to Congress that community health centers be treated as extensions of the government, thus shielding them from some of the costs and liabilities associated with doing business. Another article highlighted the "fiscal siege" resulting in a fragmented system of care for those most vulnerable populations (Hershenson, 1991, WC1). According to the article, in Yonkers, there was just one hospital accepting Medicaid patients, causing an inundation of demand and potentially contributing to higher infant death rates. Supporting the idea of a special designation for medical facilities predominantely caring for Medicaid patients, one administrator explained that "each visit cost the hospital $220, while Medicaid's reimbursement fee to the hospital had been frozen at $60 for many years." Without private insurance payments to offset these charity care cases, a public health center model is not sustainable for any specialty.

The financial challenges faced by indigent healthcare medical facilities were brought to the forefront by the hard work and dedication of community health center administrators and advocates. Their efforts to educate legislators on the problems their constituents faced, coupled with the informative media coverage, led to a significant breakthrough. Congress extended and expanded the community health center program, establishing its new designation and entitling eligible entities to access federally funded malpractice insurance, deeply discounted pharmaceutical pricing, and favorable Medicaid and Medicare reimbursement rates, marking a significant achievement in healthcare policy. The establishment of the FQHC designation was a testament to the power of advocacy and the potential for positive change in healthcare policy.

Clinton: Administrative Economy and Efficiency

From the beginning of his presidency, President William Clinton focused his domestic efforts on fighting crime, improving conditions for families, creating jobs, improving education, and investing in technology to boost the economy and support the American people. As a Democrat, he committed to ensuring equal opportunity for all citizens and providing social safety nets for those more marginalized members of our society. Although he promoted the continuation of social programming, he was also conscious of overspending and emphasized the need for economy and efficiency in social programs and initiatives. He spoke about the burden of healthcare costs in his first Congressional address, recounting the plight of a family whose medical bills led to bankruptcy. The example emphasized the need for healthcare reform, underscoring the challenges faced by constituents without insurance or with preexisting conditions. President Clinton also discussed the burden on small businesses to cover employee insurance costs. However, he also acknowledged that some societal issues were beyond the government's reach; therefore, individual responsibility and community involvement were critical to moving the nation toward a better future.

During President Clinton's administration, community health center appropriations increased by 65 percent from $757 million in 1995 to $1.2 billion in 2001. However, this increase came after the community health center program survived a significant reform effort aimed at promoting managed competition among programs. This effort resulted in the Health Center Consolidation Act of 1996, which consolidated all federal indigent healthcare efforts into a single federal agency. President Clinton's pledge to make government leaner and more efficient was the nexus of this plan, as introduced in his first address to Congress, stating:

> I will deliver to Congress a comprehensive plan for health care reform that finally will bring costs under control and provide security to all of our families, so that no one will be denied the coverage they need but so that our economic future will not be compromised either. . . . We must make choices. . . . We cannot continue these spending patterns in public or private dollars for health care for less and less and less every year. We can do better. And I will work to do better. (Clinton, 1993, para. 25–6)

During this period, the federal government tasked community health centers and other social programs with clearly demonstrating their value to validate their existence in the federal landscape. To be viewed as effective, health centers had to expand their benefits beyond traditional models of care and reach specialized populations not sufficiently served by existing healthcare structures. Media coverage reflected this new urgency to showcase the unique contributions of health centers. For instance, one article spotlighted a health center reading program that distributed books to pediatric patients and their families in the waiting room to educate parents about the importance of literacy for childhood development as a regular part of the medical visit. This innovative approach to healthcare delivery was a testament to the holistic care health centers provide. The reporter quoted a physician and reminded the reader, "If you can't read, you can't do anything and you start out at the bottom of the totem pole," reinforcing the holistic approach to healthcare offered by the program (Boyle, 1996, CN22).

Three other articles from this period focused on specialized populations uniquely served by health centers, such as individuals identifying as LGBTQIA+ and undocumented immigrant communities. In a piece about the unique healthcare needs of lesbians, the author educated the reader about the judgment and discrimination that could make the acquisition of medical care untenable for this group. The result was a population that was much more susceptible to the late detection of serious diseases (Fisher, 1998). The article emphasized the need for medical spaces that understood and met the unique needs of this population, and how health centers in New York City often served as the only medical facilities to rise up to the challenge. Similarly, two other pieces recognized the specific healthcare needs of farm workers. Traditional sources of medical care were an ineffective option for this population because of individuals' historical and justified mistrust of institutions, cultural and language differences, geographic isolation, and the presentation of occupational ailments, such as hypertension, dermatological problems, and muscular-skeletal problems. One article noted that an estimated 300,000 individuals in a jurisdiction required community health center services due to being uninsured, underinsured, or undocumented, providing a sense of the size and scope of the problem for that area (Sierpina, 1995). In addition to highlighting the need, the piece also emphasized the mushrooming costs associated with emergency care for these populations, who often lack sufficient insurance to cover primary and acute healthcare needs.

This educational campaign about the value of community health centers was successful, and by the end of President Clinton's administration, patient visits were up over 40 percent. This surge was accompanied by a higher proportion of visits with lower-cost nonphysician providers, such as nurse practitioners, case managers, and physician assistants. The Health Center Consolidation Act of 1996 propelled a renewed emphasis on continuity of care through follow-up visits and specialist referrals, as well as increased partnerships with managed-care organizations. This data highlights the effectiveness of community health centers in addressing the diverse healthcare needs of the American public, a key aspect of President Clinton's healthcare reform strategy.

Bush Jr: Stepping up to the Challenge

Like his father's tenure nearly a decade before, international conflict and domestic reform punctuated George W. Bush's time in the White House. His first Congressional address took place merely months after the September 11 attacks and highlighted the strength and unity of the nation in the face of war, recession, and terrorism. In his speech, he outlined the progress made in the war on terror, particularly in Afghanistan. He also emphasized the ongoing threat posed by terrorists and regimes seeking to develop and maintain weapons of mass destruction, asserting that while "our war on terror is well begun, but it is only begun," which foreshadowed the continued international conflict ahead (Bush, 2002, para. 27). Domestically, President Bush called for cooperation across the aisle and between all levels of government and the public and stressed the importance of strengthening homeland and economic security. Significant priorities for his administration included job creation, education, energy policy, tax cuts, welfare reform, healthcare, and retirement, with his chief domestic causes being homeland security and education outside of fighting Al-Qaeda. The No Child Left Behind effort, an initiative to create an educational social safety net catching K–12 students before they fall through the cracks, was a signature program. Although not explicitly named in his speech, indigent healthcare was a high administrative priority and received a significant boost during Bush's administration. Like his education initiatives, legislation impacting community health centers focused on creating a safety net for the healthcare system to reduce emergency room use and uncompensated care, recognizing that "economic security can vanish in an instant without health security" (Bush, 2002, para. 43).

The revised regulations required federally designated community health centers to see anyone presenting for care, regardless of their ability to pay, and to implement a sliding fee system based on an individual's family size and income. To both incentivize and subsidize these rules for community health centers, congressional reauthorization and funding nearly doubled the program's federal budget, making its appropriation more than $2 billion for the first time since inception. This significant boost in funding not only ensured the sustainability of the program but also underscored the crucial role of community health centers in reducing emergency room use and uncompensated care, thereby enhancing the overall healthcare system.

Recurring themes from the *New York Times* articles of the time included cultural competency, the use of data as evidence, barriers to care, and budget cuts. Regarding cultural competency, articles centered on the experience of patients and social groups with specialized care needs, such as gay men living with HIV/AIDS, migrant workers, or individuals living with severe chronic pain. Community health centers are uniquely capable of understanding these populations and having the knowledge and competence to effectively meet their needs in a way that traditional medical providers have fallen short. In an article about a new strain of drug-resistant HIV emerging and the need to educate patients and the larger community, one activist associated with a community health center predominately serving LGBTQIA+ individuals states, "It's been challenging to create relevant and attractive ways to talk about the gay and bisexually active community about affirming life, preventing H.I.V. in ways that are not old. The dialogue just got more complicated" (Confessore, 2005, p. B4). The article goes on to discuss the innovative ways community health centers achieve this goal through education and compassionate care. Another article discusses the difficulties of effectively treating individuals suffering from chronic pain conditions and highlights a community health center with a unique pain clinic approach rooted in holistic care. Described as a "chronic pain circle," a patient sits in a room with a diverse panel of medical providers and receives a slow, detailed, and comprehensive examination, which led by everyone simultaneously. The reporter describes, "Initial patient histories and physical exams take about an hour and a half. There are many of us asking [the patient] questions, but the atmosphere is relaxed. There are no white coats, no beepers going off, and of course the snacks help" (Dairz, 2002, p. F1). This quote provides another example of the model of care that is possible at community health centers but foreign to traditional medical providers.

There has also been an emerging emphasis on data in community health center-related articles from this timeframe. In addition to providing data about the size and scope of a highlighted issue, articles also include data as evidence of a perceived need or a solution to the problem. While qualitative narratives are still employed, the reference of data and research becomes more common in this timeframe. For example, in an article covering drug use and HIV, themes of the importance of cultural competency and the use of data as evidence are both present. The article states, "A recent survey by the San Fransico health department . . . has highlighted the issue. Of the gay respondents, 32 percent said they had used Viagra in the past year, compared with just 7 percent of the heterosexuals, and many of the gay men indicated that they had combined it with Ecstasy, methamphetamines, poppers and other recreational drugs" (Tuller, 2001, p. F5). The use of data in this article underscores the intensity of the problem of recreational drug use and sexual behaviors among gay men and points to the unique challenge of working with this population, which is best suited to knowledgeable and compassionate care providers. A final article continues these themes within the context of the experience of migrant workers. A population described as hard-working in physically challenging work environments but living in poverty with no paid days off and living in areas with limited access to doctors, the *New York Times* characterizes migrant workers as a vulnerable group ignored by traditional medical providers. An outreach worker for a migrant health program at a rural community health center explains to a reporter, "For all the difficulties inherent in their work, clinic staff members clearly relish what they do with an almost missionary zeal. . . . You couldn't pay a doctor to be here. . . . The people I see need access. They're grateful for the care." (Fuentes, 2001, p. F5).

Obama: Reinforcing the Medical Safety Net

As the nation's first Black President, President Barack Obama faced numerous challenges from the onset of his administration, including terrorism, war, economic recession, high unemployment rates, and the need for financial reform of government programs. He campaigned on the idea of change and committed to stabilizing the economy, creating jobs, and supporting small businesses in his early days in office. Fiscal responsibility, on the part of both government and the private sector, was another prominent goal of President Obama's administration. At the end of his two productive terms, President Obama was celebrated for his willingness to cooperate and reach across the aisle to propel legislation benefiting all Americans forward. He succeeded in implementing fundamental changes that improved the lives of Americans, most notably the Affordable Care Act (ACA).

In his initial address to a joint session of Congress in February 2009, he outlined the plan for the ACA, which included the community health center program as an important part. He stated in part:

> [The budget] makes the largest investment ever in preventive care, because that's one of the best ways to keep our people healthy and our costs under control. . . . It includes a historic commitment to comprehensive health care reform, a down payment on the principle that we must have quality, affordable health care for every American. (Obama, 2009, para. 40–1)

The subsequent set of healthcare bills included the American Reinvestment Act of 2009, a stimulus package to move the country out of recession. Community health centers received over $3 billion in funding for expansion, facilities, and health information technology adoption as part of this package. These funds helped community health centers prepare for the passage of the ACA a year later, which included a Medicaid eligibility expansion option for states, which would more than double the number of individuals covered by state-subsidized health insurance. Lobbyists for traditional medical providers expressed discontent with the possible influx of new Medicaid patients in their practices. To allay these fears, the ACA tapped community health centers to absorb much of this demand and provided unprecedented infrastructure and staffing growth funds to meet the anticipated demand. This brought awareness of the community health center program into the public's mainstream, solidified its position as an integral part of the US healthcare safety net, and added $11 billion to the program's then-$2 billion federal allocation. However, this growth cooled in 2010 when political control in the Senate and House shifted to the right, and there was a renewed focus on reducing the federal budget and repealing healthcare reform. The community health center program lost approximately $600 million, which slowed the expansion of the program. In fiscal year 2014, President Obama doubled down on his promise that healthcare access for all $3.7 billion is secured; fiscal years 2016–18 see an additional $8.1 billion allocation.

New York Times articles from this period underscored the need for renovated facilities and a greater number of medical professionals dedicated to primary care in medically vulnerable communities. An article from early in Obama's administration highlighted a summer program to send second-year medical students to this healthcare environment, particularly in the Northwest. The author stated, "Many [students] concluded that it was critical to reorient a reimbursement system that profoundly devalued primary care and prevention. Long hours and high patient volumes with relatively little pay." The article went on to foreground one student who believed that the effort was not worth it, stating, "There is too little incentive after going through this education and being in school for the majority of your life to want to pursue a career just in primary care" (Sack, 2009, p. A18). However, many other students were quoted as becoming more committed to helping communities in need as a result of their volunteer experience. Overall, the article educated readers about the struggles many new doctors face when deciding where to contribute their talents and the need for funding to attract and retain them in less-resourced communities.

Similarly, a piece from the middle of Obama's administration described a health center as "hope in the wreckage," especially for rural populations in a broken healthcare system (Hansen, 2012, p. SM22). The author emphasized that insurance and physical access to clinics in a disjointed system are insufficient to change severely depressed areas. Focusing on such communities of Mississippi, the article pointed out that "millions of dollars pour into the state every year . . . but the lives of Mississippians do not improve." Statistics and descriptive stories conveyed the dire need for a medical safety net prioritizing holistic care. Among these were families shopping at gas stations for weekly groceries, skyrocketing HIV rates, and Black infant mortality rates twice that of white infants, costing the state millions of dollars in neonatal units. Advocating an innovative approach to preventive healthcare delivery for medically vulnerable patients, a clinical care worker stated, "The hospitals need a third party trained to discern what exactly will help a patient, and that party must come from the patient's world: talk the same, share similar fears and frustrations and life experiences" (Hansen, 2012, p. SM22). A plausible message is that the modern focus on healthcare efficiency has caused health centers to evolve into more conventional medical businesses. ACA funds could help them return to their roots. By focusing on the social and environmental determinants of health, centers could transform communities while maintaining a high standard of care. A third article highlighted health centers that helped revitalize communities by repurposing brownfield sites, specifically "taking a potential source of health problems for the community and turning it into a place for health care" (Kaysen, 2012, p. B8). Again, pressures to meet growing patient demand while doing more with less were a central theme of the *New York Times*' coverage.

Trump's First Term: Defense and Derailment

Punctuated by themes of infamy, conflict, and polarization, President Donald Trump's tenure in the White House starkly contrasted with President Obama's efforts to unite the nation. A businessman and showman at heart, President Trump won the election by promising to bring the United States back to a place of traditional stability and power in the world. Outside of personal and professional controversies, history will associate President Trump's administration with the Covid-19 pandemic and its response, including several trillion dollars spent on financial relief, testing, and vaccinations. In his first joint address to Congress years before the arrival of Covid-19 in the United States, he emphasized the strength of the American people and his plans to address domestic challenges such as infrastructure, drug pricing, trade deals, and immigration reform to strengthen the nation further. He also emphasized the need for more robust border security and "immigration policies that focus on the best interests of American workers and American families" (Trump, 2018, para. 37). President Trump was also committed to dismantling what he called "disastrous Obamacare" and was successful in the repeal of the individual mandate (Trump, 2018, para. 15).

President Trump's administration presented a challenging landscape for community health centers, with the administration's disapproval of the ACA evident. However, community health centers, demonstrating their remarkable resilience, continued to showcase their value. In January 2019, they were the focus of a full committee hearing about the benefits of the community health center program titled "Access to Care: Health Centers and Providers in Underserved Communities." Witnesses at the US Senate hearing underscored the crucial role of the healthcare program in providing essential healthcare to underserved communities and the necessity of stable and full funding to address lapses that impact operations. The discussion also highlighted the role of community health centers in training providers for rural areas and providing employment opportunities for healthcare professionals in vulnerable communities. The hearing reaffirmed the importance of primary care, reducing unnecessary healthcare costs, and expanding the healthcare workforce, providing reassurance about the future of healthcare. A year later, the Covid-19 pandemic began.

During the pandemic, the community health center program once again rose to the occasion, becoming a primary source of Covid testing and vaccinations for Medicaid and uninsured populations. Given the novelty, urgency, and scope of the Covid-19 response, this was an immense undertaking. However, the challenges were significant. Most centers experienced a loss in revenue as they canceled routine follow-up appointments and conducted urgent appointments via telemedicine, which had significantly lower insurance reimbursement rates than face-to-face visits. Consequently, many centers had to close, and for those that could remain open, a significant portion of medical and support staff were laid off or fired to keep the organizations afloat.

Media coverage at this time overwhelmingly focused on the fear and confusion surrounding the Covid-19 pandemic and its adverse effects on community health center operations and budgets. An article entitled “Community Health Centers, Lifelines to the Poor, Are Forced to Cut Back” illustrated this trend, highlighting a center that had laid off more than a third of its staff due to the pandemic (Johnson & Goodnough, 2020, p A17). The authors warned that the nation’s system of community health centers could “collapse” as “the centers are being battered by ferocious crosswinds as routine medical care—including dental work, a primary source of revenue—has been put on hold to conserve resources and keep the coronavirus from spreading.” The article also points out that while community health centers will get a $1.3 billion boost from President Trump’s stimulus package, that amount will be shared across all 1,362 centers, covering just over one month of expenses. Another article pointed out that community health centers are confused about their role in the fight against the spread of the virus, stating that “some community clinics feel they are caught in the middle of a haphazardly planned vaccine rollout” (Goldstein, 2021, A5).

Coverage from the early years of Trump’s administration focused on the plight of undocumented immigrants who felt under attack after President Trump’s ascension to the White House. There was a palatable fear of deportation for many, resulting in individuals staying inside their homes even when they needed medical care. One reporter explains, “Shaken by the Trump administration’s board deportation orders, they and many others like them are retreating into the shadows, forgoing screenings, medications and other essential medical care” (Hoffman, 2017, p. D1). Another article reports, “Community urgent-care clinics have had patients asking to have their records removed, and some immigrants have refused to sign up for food assistance programs, citing worries about deportation and family separation” (Nguyen & Hirota, 2018, p. A27). These two broad examples, rooted in fear of the unknown, illustrate the intensity of the sociopolitical environment within which the community health center program had to navigate and stay afloat.

It was not until President Biden's administration that administrators could return to a familiar landscape of promoting hope, growth, and comprehensive patient care.

Biden: Shoring Up the Gaps in the Net

By the time President Biden came into office in 2021, the disruption and fear of the pandemic had been replaced by a new rhythm of American life full of social distancing, masks, nasal swabs, and vaccinations working to contain the spread of disease. Unsurprisingly, Covid-19 and a return to normalcy were primary themes in his first Congressional address, as he acknowledges, "For more than two years, COVID has impacted every decision in our lives and the life of this Nation. And I know you're tired, frustrated, and exhausted. That doesn't even count the close to a million people who sit at a dining room table or a kitchen table and look at an empty chair because they lost somebody" (Biden, 2022, para. 77). Outside of the pandemic, President Biden also highlighted foreign challenges such as the Russian invasion of Ukraine and the global response, as well as domestic issues like the economy, infrastructure, and inflation, which was a primary domestic concern as the country emerged from the pandemic. President Biden's primary defense to this threat was the American Rescue Plan. Emphasizing the bill's importance, he states, "And with all the bright spots in our economy—record job growth, higher wages—too many families are struggling to keep up with their bills. Inflation is robbing them of gains they thought otherwise they would be able to feel. I get it. That's why my top priority is getting prices under control" (Biden, 2022, para. 47).

The American Rescue Plan Act of 2021, a comprehensive law passed by the US Congress, was crucial in providing relief and support during the Covid-19 pandemic. It included provisions for multiple areas of support to individuals, communities, and businesses, such as premium tax credits, Medicaid expansion, education support, small business assistance, and housing cost relief. Importantly, the Act allocated an additional $7.6 billion to community health centers, supplementing their regular grants and ACA funding. This significant boost was aimed at offsetting the disastrous impact of Covid-19 on their operations and budgets. Despite the challenges, community health centers, serving as the frontline of the pandemic response for Medicaid and uninsured populations, demonstrated remarkable resilience. They were able to solidify their critical role in shoring up the healthcare safety net, rehiring staff and providers, improving infrastructure, and purchasing the necessary resources for combatting Covid-19 in their communities.

New York Times articles from this period returned to covering the unique role community health centers play in providing culturally competent services to marginalized groups and communities. It is interesting to note, however, that the groups highlighted in these more contemporary pieces include smaller ethnic groups, such as Asian and Pacific Islanders, or occupationally defined groups, such as coal workers, as opposed to the traditional pieces focusing on individuals living in poverty, immigrants, and Black communities. These articles highlight the vital role of community health centers in providing culturally competent services and the importance of these institutions in addressing health disparities. One article discusses how community health centers serving a significant proportion of Asian-American patients have been "inundated with requests from companies and organizations that wanted to host support groups for their employees" following the anti-Asian backlash associated with the pandemic, resulting in depression and anxiety in individuals (Lukpat, 2021, p. 17). Another article points out that one of Nancy Pelosi's guests at President Biden's first joint address to Congress was a doctor who runs a community health center primarily serving Asian-Americans and Pacific Islanders, emphasizing the attention to these populations on a federal level (Broadwater, 2021). An article from early 2022 discussed the pressures and burnout from Covid-19 on health center workers and foregrounded a community health center worker providing services to Indigenous populations in California. She states, "As an indigenous woman, I'm just trying to do my best to help my people and others like us survive this pandemic. I'm doing good work in a community health center that tries to help the most vulnerable. And even if I get death threats or scorn and yelled at by those who misunderstand what we're doing, my family sees me. And loves me" (Kalamangla, 2022, para. 18). A final article underscores the challenges of immunizing Hispanic populations against Covid-19 due to technology and language barriers, in addition to a mistrust of government. The reporter notes that community health centers have been at the forefront of this effort. The article cites data from the Kaiser Family Foundation, noting that

> federal efforts to close the vaccination gap for Hispanic Americans through community health centers, while limited in size, are having some success. While the Hispanic share of the U.S. population is about 18 percent, Hispanic people made up more than a quarter of those nationwide who received their first dose at a community health center. (Walker, 2021, p. 6)

Outside of the pandemic, community health centers were also highlighted with their work with communities suffering from opioid use disorder, a condition receiving a lot of federal and state attention prior to the pandemic. One piece points out that opioid drug overdoses and deaths actually spiked during the pandemic, and community health centers are at the frontline of getting the situation back under control, especially in rural areas (Weiland, 2022). Another piece focuses on the community health center's role in helping native Hawaiians find solace after the deadly 2023 fires. A community health center is highlighted for providing a space of spiritual healing for the community, which "over the course of 10 days started out drawing a few dozen people and grew to as many as 100 in person and more than 80,00 watching a livestream on social media. Oprah Winfrey, who has a home on the island, attended quietly on the last day" (Graham & Cave, 2023, para. 9).

Conclusion

The thematic analysis of Presidential agendas, legislative initiatives, and *New York Times* articles about the impact and role of community health centers in American society illustrates a significant narrative shift for the program. To maintain favorable funding and legislative support, community health center administrators and advocates have carefully curated the program's image across political transitions over the past sixty years since the beginning of the War on Poverty through President Biden's administration. Community health center administrators and advocates have found an invaluable ally in the media in their struggle to adapt to the sociopolitical forces that shape public perception of poverty and indigent healthcare during this period. Traditional media outlets have played a pivotal role in raising public awareness about the human experience of poverty and inequality and telling the story of community health centers within this context. The *New York Times*, while not the sole representative of US media coverage of health programs for the marginalized, serves as a powerful example of how socially responsible coverage can remind the public of its moral, economic, and social responsibilities. The strong ties between the *New York Times* and health center advocates over the past sixty years have helped keep community health centers on the public agenda within a policy subsystem that tends to overlook the medical needs of marginalized populations. The coverage built and supported by the *New York Times,* in concert with community advocates, humanized the issue of health disparities. It has consistently highlighted the community health center program as a critical component of the healthcare safety net over several reauthorization periods. The administrative dexterity of health center leaders and advocates has made the program relatively easy to defend; it offers demonstrated benefits to communities in crisis and cost savings to taxpayers. The media have been indispensable in keeping the national public aware of these benefits, especially during times of policy change.

The analysis presented in this chapter highlights the juxtaposition between the traditional perception of community health centers as addressing the healthcare demands of historically marginalized communities and thematic policy agendas characterizing selected presidential terms. It supports the assertion that public administrators kept pace with the changing agendas of successive presidential administrations and public opinion, morphing the narratives of poverty and healthcare for marginalized populations to match contemporary definitions and contexts, ultimately fortifying the program's social meaning since its inception. This approach has been a delicate dance, as mission creep is always risky when shaping administrative goals to match external expectations. While broadening an organization's mission to address emerging needs or capitalize on new opportunities to build legislative alliances can be tempting, mission creep can carry significant risks. It can lead to a loss of focus and dilution of resources as administrators stretch to meet extraneous expectations and react to environmental threats, potentially compromising the organization's core competence and effectiveness.

Additionally, shifting narratives can confuse stakeholders such as patients, potential patients, and supporters and erode trust in the organization's ability to deliver on its promises. Perception matters, but so does staying true to oneself, even in the case of organizational leadership. Therefore, it is essential to strike a balance between responding to changing circumstances and staying true to the organization's original mission and values.

Community health center administrators have managed this threat beautifully, although it has not been without sacrifice. While many community health centers are integral partners in their communities, the anti-poverty spirit of strengthening neighborhoods has become less of a focus. As a result, improvements in the health of community health center patients are well documented. However, widespread reductions in health disparities between target populations and the rest of society remain elusive. Federal projects, such as Healthy People 2030, seek to close these gaps with community health centers serving as critical partners. However, these goals are almost secondary to the perceived and actual expectation that community health centers primarily serve to fortify the larger healthcare system and reduce the costs of uncompensated care for emergency rooms and hospitals. Center administrators must balance this need without full financial and operational support for neighborhood care, which is a challenging task. Nonetheless, administrators persist and do more with less to simultaneously meet the funder's expectations on the federal level and equity needs on the neighborhood level. This is why community health center administrators have consistently operated at a level of excellence and enjoyed bipartisan congressional and Presidential support during nearly every authorization period since the program's inception. For this, they should be applauded and recognized despite the challenging environments in which they function, as illustrated in this analysis.

4

Stubbed Toes and Mixed Signals

Unintended Consequences of Well-Intentioned Administrative Strategy

Researchers and community health advocates have long touted considerable cross-sectional and longitudinal evidence of the positive effects of community health centers on users compared to similar non-users. In study after study, community health center patients are reported to enjoy better quality of care, access, care continuity, and fewer racial and ethnic disparities in health measures as compared with their noncommunity health center user counterparts (Dor et al., 2008; McMorrow & Zuckerman, 2014; Kirby & Sharma, 2017; Saloner & Levin, 2020). The significant role of community health centers in improving health outcomes instills confidence in their effectiveness. Community health center patients exhibit more positive indicators of health, better access to care, and higher satisfaction than patients with similar personal characteristics and receiving care at traditional medical facilities, which do little to address the social needs of their patients (Politzer et al., 2001; Shi et al., 2004; Strelnick, 2004; Hicks et al., 2006; Dor et al., 2008). In one critical longitudinal study, Bailey and Goodman-Bacon (2015) found that the emergence of community health centers before 1970 significantly correlated with lower age-adjusted mortality rates fifteen years later. This outcome was primarily due to preventative health-related improvements in cardiovascular health measures for community health center users over fifty compared to their non-user counterparts.

Outcome variables such as patient satisfaction rates, evidence of standardized care plans, preventative screening rates, and documented improvements in chronic disease measures are standard methods for measuring community health center efficacy. In a typical study, O'Malley et al. (2005)

assess community health center effectiveness based on the number of hospitalizations and emergency department visits for ambulatory care for community health center patients compared to similar patients seeking care in noncommunity health center practices. The high patient satisfaction rates in community health centers provide reassurance about the quality of care provided. Similarly, in their meta-analysis of community health center efficacy studies, Bailey and Danziger (2013) argue that "evaluations of community health centers on measures of quality of care, improving access, reducing disparities, and cost-effectiveness are overwhelmingly favorable" when comparing efficacy measures such as the percent uninsured patients served, the elimination of racial disparities in access, and decreased infant mortality and increase life expectancy for at-risk populations (2013, p. 249).

As discussed in Chapter 3, advocates often point to these positives when persuading legislators about the benefits of community health center budgetary support across several presidential administrations. Over the past few decades, community health centers have carefully positioned themselves as critical partners in the control and reduction of uncompensated costs to the healthcare system, such as those resulting from unnecessary emergency department visits, by absorbing potential uncompensated ambulatory care visits for Medicaid, uninsured, and underinsured individuals (Chen et al., 2015; Hutchison et al., 2018). This absorption is especially critical for historically marginalized populations with chronic medical conditions, as there is an association between barrier-free access to healthcare and the increased likelihood that individuals from this population receive critical preventative care, reducing the need for avoidable emergency department visits (DeVoe et al., 2003; Liu, 2007; Taylor & Salyakina, 2019). Evidence also suggests that although community health center penetration (e.g., the proportion of a target population using health center services) is relatively low in medically underserved neighborhoods, incremental increases in penetration rates are associated with the improvement of select health-related outcomes for specific populations within the service area (Shin & Rosenbaum, 2003).

Despite the consistency of their positive findings, community health center efficacy studies also share some methodological limitations. First, a common limitation of community health center efficacy studies is that the studies measure the same health outcomes that require reporting to the federal government, such as the administration of flu shots and Pap smears (Politzer et al., 2003; Shi, 2009). While these types of studies demonstrate evidence of the positive effects of being a community health center patient, it is unclear whether these findings can be extrapolated to other indicators of health, such as self-reported health status or the prevalence of pain, which are not widely tracked or prioritized by the community health center program rules. For instance, low birth weight and immunization rates are core pediatric outcomes tracked on the Uniform Data Systems report, an annual demographic and performance report submitted by community health center administrators to the federal government. In their efficacy study, Politzer et al. (2003) used low birth weight as one of the criterion for positive health outcomes in pediatric patients. More specifically, they found that upon giving birth, Black community health center prenatal care patients delivered low birth rate babies at a rate 20 percent below the national average for Black mothers, a significant feather in the hat for the community health centers. Similarly, Regan et al. (2003) reported evidence of higher immunization rates among children receiving care at community health centers as compared to children of similar backgrounds nationally who did not receive care at a community health center. However, they were still below the expected range for all Black patients and those with Medicaid insurance. These two examples are just a sample of many that illustrate that while these studies are essential for community health centers and those that use them, expanding definitions of success to include measures outside the required reporting standards for the community health center program is also important. Broader definitions more accurately reflect the multitude of benefits community health centers bring to their patients and neighborhoods and can highlight areas where more support, attention, and resources are needed.

Second, many widely cited studies do not adjust for the likelihood of a fundamental difference between patients seeking medical care at community health centers and patients who do not (Politzer, 2001; Shi, 2004; Shi, 2009). In one prominent example, Politzer et al. (2001) compared National Health Interview Survey responses of uninsured community health center patients to uninsured noncommunity health center patients and found that community health center patients were half as likely to delay or forego care or choose not to fill prescriptions due to cost. They also found that the rate of diagnostic cancer screenings and healthy lifestyle counseling for uninsured community health center patients was higher than the national rate. Similarly, Shi et al. (2009) compared selected health indicators for randomly selected community health center patients with a similar group of noncommunity health center patients and patients without a regular source of care, and found that community health center patients had higher utilization of healthcare services than the comparison groups. Moreover, racial and ethnic disparities in access and quality of care measures all but disappeared for health center patients as compared to comparable non-users. In both examples, the demographic and personal characteristics of regular community health center users may be distinct from those of private doctors and traditional facilities or those who do not seek care. The difference in health outcomes between the two groups reflects a measure of the inter-group difference, rather than the effect of the community health center itself on health outcomes. Fundamental social differences between those who choose to use community health centers and those who do not include levels of social capital, stigma, cultural expectations, level of trust, and transportation inequities, which can make these types of efficacy studies susceptible to self-selection bias. Comparing users' health measures and outcomes to those of non-users without considering these social nuances could lead to overestimating the benefits of community health center usage, especially for medically vulnerable populations.

Moreover, when considering Shi et al.'s (2009) study of disparity reductions in community health center patients, one of the authors' main arguments is that a 2-percentage-point disparity in insurance rates between the "near-poor" and "poor" patient groups at noncommunity health center facilities disappears for community health center facilities. This difference led to the conclusion that community health centers likely mitigate the effects of income inequality for their patients as a function of their patient-centered and holistic model of care. At the same time, it is feasible that this outcome is due to programmatic and operational differences between community health centers and traditional medical providers. It is also possible that higher-income patients with insurance are more likely to utilize traditional medical providers and facilities than uninsured or underinsured individuals. This is because while community health centers aspire to be primary care providers of choice, the public often perceives them as providers of last resort or primarily serving special populations. In their 2015 paper, Bailey and Goodman-Bacon discuss this potential complicating factor, finding that it is reasonable to conclude that the community health center population may have a lower percentage of higher-income insured patients, making the rate statistically indistinguishable from the percentage of insured lower-income patients and eliminating the disparity between income groups for the community health center facilities.

A New Approach to Evaluating Community Health Centers

To more fully understand the broad impact of community health center access on the collective health of neighborhoods as initially intended, it is necessary to place the contributions of the community health center program within the larger social and political context in which it functions. Doing so necessitates exploring and accepting alternative definitions of health that may fall outside the traditional scope of mandated reporting, as well as and assessing the impact of having access to a community health center for patients and non-patients alike. Due to low penetration rates, thousands of individuals reside in and around neighborhoods where community health centers are located but choose not to be regular users of the services. Understanding how they are impacted by having a community health center in their neighborhood is another essential perspective from which to assess the effect of community health centers on populations. As explained in the previous chapter, the federal community health center program is one of many initiatives that aim to improve outcomes for historically marginalized groups in the United States. It is, therefore, quite possible that the population-level improvements in health disparities are less attributable to the impact of community health centers and more of a result of broader political and environmental changes, such as legislative changes in healthcare, economic policy, and the combined efforts of public policy programs in general, which may improve health outcomes for community health center and those in the area but choose not to use health centers. Parsing out this distinction is vital for understanding the isolated effect of having a community health center in one's neighborhood.

Bailey and Danziger (2013) argue that "although clearly the health of the poor improved dramatically since the War on Poverty . . . eliminating disparities is far more difficult, because health insurance and access to care are only two elements in influencing health; knowledge, resources, health-related behaviors, and security purchased through higher incomes are clearly other important determinants" (p. 260). The community health center program's risk-based approach to reducing health disparities has documented benefits for compliant and regular users of services. However, it does little to eliminate disparities on a larger scale, especially if the proportion of users in an area is small. Overcoming this hurdle requires a multipronged approach to strengthening and expanding the safety net while leveraging existing resources to draw more patients into care and return to the systematic focus on community building—an original goal of the program. The first few steps in this evaluative process include understanding who benefits the most from the programs, identifying groups that have access but do not show measurable benefits, and recognizing the social barriers contributing to this difference. As we know, individuals are not randomly distributed across geographic landscapes. Those living in highly marginalized urban areas are much more likely to have access to a community health center but start with lower levels of health.

In contrast, those living in more affluent areas are less likely to have geographic access to a community health center, but they tend to have better health outcomes. Due to existing structures of social inequality, those occupying spaces at lower rungs of the social stratification ladder are exposed to more negative risk factors that influence health outcomes than those experiencing life at higher levels (Johnson, 2018). In traditional healthcare systems, historically advantaged groups tend to have better physical, financial, and cultural access to care. At the same time, those at the bottom of the social stratification are more likely to receive fractured care littered with various barriers, including racism and the denial of basic human entitlements (Farmer, 2010). These barriers often manifest as a lack of appropriate insurance, financial hardships, transportation issues, language barriers, and dispassionate care, which ultimately impedes their ability to get the medical help they need promptly (Carillo et al., 2011; Tarraf et al., 2014).

Optimal assessments of community health center efficacy will consider the mediating effect of self-selection by controlling for nuanced social factors before evaluating the primary relationship between access and health outcomes. They will also use measures of health beyond those required by federal reports. In this context, a distinction between users and non-users is a crucial consideration. My 2023 study, which compares individuals with proximal geographic access to a community health center to those without geographic access, regardless of usage level, aims to fill this gap, offering a unique perspective on the isolated effect of having a community health center in one's community (Gadson, 2023). The study findings reveal that while individuals living in close spatial proximity to community health centers are more likely to identify a community health center as a regular source of care compared to those without proximal access, the effect of community health center access on the identification of any source of regular healthcare was generally insignificant or negative, except for populations with a chronic medical condition. A brief description of this study follows, serving as a compelling example of how reevaluating community health centers can help them return to their anti-poverty roots of community transformation through improved healthcare access.

Study: Self-Reported Health and Geographic Access to a Community Health Center

Few studies isolate the effect of geographical center availability and health, particularly when considering those living in the catchment area but are not regular users. This realization is an important distinction as the primary charge of the community health center program is to provide a safety net for entire communities, not just those that actively seek out and comply with the services. Moreover, far more individuals in vulnerable communities have access to community health centers than regular users, as the effect of safety net expansion disproportionately targets historically marginalized social groups. For example,

Blacks and Hispanics are more likely to visit community health centers, and federal reporting shows that these groups are overrepresented in community health center program patient populations, as compared with non-Hispanic Whites (Hadley et al., 2006). My work takes an alternative approach aiming to quantify the effects of geographic access to community health centers on the likelihood of an individual having a regular source of healthcare. Utilizing survey and location data, the analysis employs bivariate crosstabulation with chi-square and multinominal logistic regression to quantify the relationship between variables. While not a direct proxy for health, having a regular source of care is a rough measure of one's perceived and actual access to preventative screenings and services, reducing the dependence on emergency care in the future. Community health center patients are more likely to report having a regular source of care than similar individuals who do not utilize community health centers nationally, an essential precursor for decreasing barriers to care (Politzer et al., 2001; Xu, 2002). However, it is important to note that the relationship between having a usual source of regular care and subsequent emergency department use for nonemergent conditions is nuanced in the literature. While evidence shows that those who solely use a hospital emergency department for ambulatory care are less likely to have a regular source of care, individuals with high emergency department use also tend to be high users of primary care and other hospital services (Merzel & Moon-Howard, 2002). This can be especially common for "super-utilizers," characterized as at-risk or vulnerable individuals with chronic medical conditions living in areas of high disadvantage (Byrne et al., 2003; Miller et al., 2013).

There is also mixed evidence of whether the type of primary care service (i.e., specific doctor versus facility) is a factor in predicting emergency department utilization. An early study isolating the effects of having a specific doctor versus a facility as a usual source of care suggests that the distinction makes little to no difference in the frequency of emergency department use, as facility use is a primary entry point into the healthcare system, especially for at-risk individuals. However, the distinction of whether a specific doctor or facility serves as an individual's regular source of care may have a differential impact on the type of primary care services offered. Having a specific doctor influences the provision of "discretionary" screenings such as blood pressure and cholesterol screening. However, regular facility use can affect the receipt of routine preventive services such as flu shots, Pap smears, and mammograms. More recent evidence suggests that individuals with no source of care or a facility as their usual source of care tend to have higher emergency department utilization as compared with those naming a specific doctor, as facility use as a regular source of care is associated with lower socioeconomic status, thus more complicated health needs (Liaw et al., 2018).

Due to the complexity and prevalence of the barriers to healthcare for vulnerable and at-risk populations, improving access to a regular source of care does not guarantee positive outcomes for these groups (Rosano et al., 2013). With the incremental improvement of any combination of these factors, the likelihood of better health outcomes may show some improvement. However, the magnitude and combination of personal and neighborhood disadvantage can have a devastating effect that is not easily ameliorated by social programs. Moreover, there is a lack of awareness amongst target populations about medical safety-net initiatives; of those that are aware, few tend to become regular users of available services, citing issues of perceived lack of need or eligibility, issues of convenience, and stigma of being a user of the service providers (May et al., 2004).

These factors make it critical to understand the relationship between spatial proximity to a community health center, the likelihood that target individuals will adopt the facility as a regular source of care when evaluating the program, and how social and structural factors influence this primary relationship. To return to its anti-poverty roots and fundamentally change the root causes of race and class-based health disparities in the United States, the community health center program must expand its current and safe position as a safety-net health provider, ensuring cost reductions for the entire healthcare system, and prioritize structural solutions that transform community health outcomes (Martinez, 2019).

Methods

The methodology employed in this study is robust and reliable. To quantify the effect of spatial proximity to a community health center on the likelihood of having a usual source of care as a proxy for quality healthcare access, this study employed statistical models using bivariate crosstabulation with chi-square and multinominal logistic regression utilizing secondary data sourced from the 2018 Southeastern Pennsylvania Household Health Survey (SPHHS). A discontinued report formally administered by the Public Health Management Corporation in Philadelphia, the SPHHS is a bi-annual, randomized telephone survey of adults that collects data on various health behaviors and outcomes for households in the Philadelphia metropolitan area (Public Health Management Corporation, 2024). Due to funding constraints, the survey was put on indefinite hold in 2019.

The concept of having a usual source of care, a measure of perceived healthcare access, was operationalized through the combination of two SPHHS questions: (1) "Is there one person or place you usually go to when you are sick or want advice about your health?" and (2) "When you go to the place you go most often, where do you usually go—to a private doctor's office, a community health center or public clinic, a hospital outpatient clinic, a hospital emergency room, or some other place?" Responses were combined and collapsed into a single variable with four response categories: None, Community Health Center or Public Clinic, Emergency Room, or Private/Other.

Community Health Center Access, the primary predictor in the model, was also created from secondary data. Community Health Center location information from the federal Health Resources & Services Administration (HRSA) website, a public, web-based source of information about federal programs providing services under the Department of Health and Human Services umbrella, was matched against respondent zip code data in the SPHHS to create the neighborhood-level variable. Details of this process and the control variable details are provided in the Methodological Appendix.

Results

This analysis employed bivariate crosstabulation and multinomial logistic regression to explore the effect of spatial proximity to a community health center on the likelihood of reporting that an individual has a regular source of healthcare. After all recoding and listwise deletion, the sample size was 7,226. The univariate statistics are available below in Table 4.1.

Table 4.1 Health Center Access and Regular Source of Care Respondent Demographics

	#	%
Source of Regular Care		
Community Health Center	332	4.59
Emergency Department or Hospital Clinic	484	6.70
Private Doctor or Convenient Care Clinic	5,729	79.28
No Regular Source	681	9.42
Spatial Proximity to a Community Health Center		
Yes	2,253	31.18
No	4,973	68.82
Race/Ethnicity		
Black (Not Hispanic)	1,364	18.88
Hispanic/Latino	337	4.66
White (Not Hispanic)	5,024	69.53
Other (Not Hispanic)	501	6.93
Sex		
Female	4,281	59.24
Male	2,945	40.76
Living in Poverty		
Yes	2,065	28.58
No	5,161	71.42
Living in Urban County		
Yes	2,806	38.83
No	4,420	61.17
Diagnosed with one or more Chronic Medical Conditions		
Yes	4,277	59.19
No	2,949	40.81
Insurance Status		
Uninsured	457	6.32
Insured	6,769	93.68
Highest Level of Education		
High School	2,143	29.66
Beyond High School	5,083	70.34
Respondent Age		
18–44	1,236	17.10
45–64	3,000	41.52
65+	2,990	41.38
N = 7,226		

The respondents were majority female (59.2 percent), forty-five years old or older (82.9 percent), and had an education level beyond high school (70.3 percent). Most respondents also had health insurance (93.7 percent) and one or more chronic medical conditions such as asthma, diabetes, arthritis, or hypertension (59.2 percent). Regarding racial/ethnic identity, most respondents identified as white, not Hispanic (69.5 percent), followed by Black, not Hispanic (18.9 percent). Respondents also generally lived above the calculated poverty level (71.4 percent), in a suburban county (61.2 percent), and in a zip code without spatial proximity to a community health center (68.8 percent). Finally, 79.3 percent of respondents reported using a private doctor or convenient care clinic as their regular source of healthcare, 6.7 percent reported using an emergency department or hospital clinic as their regular source of care, 4.6 percent reported using a community health center, and 9.4 percent reported not having any regular source of healthcare.

Table 4.2 shows the bivariate relationship between having spatial proximity to a community health center and the likelihood of reporting a specific source of regular care.

The findings of this study have significant implications for healthcare access. Those with spatial proximity to a community health center were more likely to report using a community health center as their regular source of care than those without spatial proximity to a community health center by a margin of 8 percentage points (10.1 percent vs 2.1 percent, respectively). This relationship was statistically significant at an alpha of 0.001. There was, however, little difference in the percentage reporting having no regular source of care when comparing those with and without spatial proximity to a community health center. Moreover, those with spatial proximity to a community health center were more likely to identify an emergency department or hospital clinic as their regular source of care (12.2 percent) as compared with those without spatial proximity to a community health center (4.2 percent) and less likely to use private physicians and commercial urgent care centers (68.0 percent) as compared with those without spatial proximity to a community health center (84.4 percent). The bivariate analysis suggests that those with spatial proximity to a community health center are

Table 4.2 Likelihood of Reporting Specific Regular Locations for Care by Proximity

	Spatial Proximity to a Community Health Center (CHC)	
Source of Regular Care	**Yes (%)**	**No (%)**
Community Health Center	10.12	2.09
Emergency Department or Hospital Clinic	12.21	4.20
Private Doctor or Convenient Care Clinic	68.04	84.38
No Regular Source	9.63	9.33

N = 7,226
$chi^2(3) = 418.1238$, $p < 0.001$

more likely to name it a regular source of care than those without access to a community health center. This is a significant finding, highlighting the success of the community health center program in reducing disparities as an important precursor to better health.

Table 4.3 details the effect of spatial proximity to a community health center on the likelihood of having specific sources of regular healthcare, controlling for personal disadvantage using multinomial logistic regression.

The first column displays the relative risk ratio of identifying a community health center as a regular source of healthcare compared to reporting having no regular source of care. The second and third columns offer similar data, comparing the relative risk ratio of an individual identifying hospital emergency departments and outpatient clinics, or private physicians and facilities as their regular source of healthcare, compared with those who report not having a regular source of care.

The effect of spatial proximity to a community health center on the likelihood of reporting specific sources of regular care was captured in three ways: the main effect of spatial proximity to a community health center, the interactive effect of spatial proximity to a community health center, and the prevalence of a chronic medical condition, and the interactive effect of spatial proximity a to a community health center and being Hispanic. The outcome variable delineated three categories for sources of care: community health centers, emergency department/hospital outpatient clinics (Emergency), and

Table 4.3 Multinomial Log Odds for the Likelihood of Reporting Specific Locations for Regular Care as Compared to Having No Regular Source of Care by Community Health Center (CHC) Proximity and Personal Characteristics

	CHC versus No Regular Source	Emergency Department or Hospital Clinic versus No Regular Source	Private Doctor or Convenient Care Clinic versus No Regular Source
CHC Proximity			
Spatial Proximity to a CHC (No Access)	0.1964	0.1416	−0.3202 *
Spatial Proximity to a CHC*Chronic Condition	0.6098 *	0.2525	0.2063
Spatial Proximity to a CHC*Hispanic/Latino	0.4722	−1.5151 **	−0.6142
Personal Characteristics			
Black (White Not Hispanic)	1.3272 **	1.418 **	0.186
Hispanic/Latino (White Not Hispanic)	0.9537 *	1.5733 **	0.245
Other Race (White Not Hispanic)	0.7651 *	1.0323 **	−0.0535
Living in Poverty (Not Living in Poverty)	0.6328 **	0.2551	−0.2072 *
Urban (Suburban)	0.3935 *	0.5263 **	0.1205
Chronic Condition (No Chronic Condition)	0.1873	0.4003 *	0.3224 **
Uninsured (Insured)	−0.1639	−0.4442 *	−0.9845 **
Control Variables			
Age	0.0279	0.0823 **	0.1611 **
Female (Male)	0.1564	0.2892 *	0.4043 **
High School or Less (More than Highschool)	0.3703 *	0.0602	−0.2814 **
Constant	−2.555	−2.231	0.9644
R^2	0.1088		
N= 7,226			
*=p < 0.05 **=p < 0.01			

private doctors, including convenient care clinics (Private). The coefficients represented the relative risk ratios of reporting each location compared with the likelihood of reporting having no regular source of care (None).

In the model, the main effect of spatial proximity to a community health center was insignificant when assessing the community health center versus None and the Emergency versus None comparisons and had a significant but negative effect on the Private versus None comparison, more specifically, when the respondent had spatial proximity to community health center, the multinomial log-odds for indicating a private doctor or convenient care clinic as their regular source of care versus no source at all decreased by 0.3202 ($p < 0.05$) holding all other variables constant.

However, while the main effect of spatial proximity to a community health center was generally insignificant or negative, there was a significant interaction effect of spatial proximity to a community health center for those with a chronic medical health condition for the community health center versus None comparison, and for Hispanic respondents with spatial proximity to a community health center in the Private versus None comparison. For individuals with a chronic medical condition and spatial proximity to a community health center, the multinomial log-odds for indicating the use of a community health center as a regular source of care as compared with reporting no regular source of care at all increased by 0.6098 ($p < 0.05$), holding all other variables constant. For Hispanics with spatial proximity to a community health center, the multinomial log-odds for identifying a private doctor or facility as their regular source of care compared with reporting no regular source decreased by 1.5151 ($p < 0.01$), holding all other variables constant. Including the interaction terms slightly improved the model's fit as measured by McFadden's R^2 statistic by 0.2 percentage points (0.1088 versus 0.1069 without including the interaction terms).

Finally, when assessing the likelihood of identifying a community health center as a regular source of care as compared with having no source of care, community health center use is positively associated with predictors of social disadvantage such as being a racial/ethnic minority, living in poverty, residing in an urban environment, and having a level of education of high school or less.

These findings suggest that individuals with spatial proximity to a community health center are more likely to utilize emergency departments and hospital clinics as a usual source of care versus having none as compared with those without spatial proximity to a community health center and less likely to regularly use a private doctor or facility versus reporting having no regular source of care. It is interesting to note, however, that this relationship does not hold for Hispanics with spatial proximity to a community health center, as this group is less likely to use private or emergency department services versus reporting no regular source of care as compared with those lacking spatial proximity to a community health center. This may be attributed to a reluctance to utilize traditional healthcare services due to issues of trust and cultural and language barriers. This insight into the role of cultural and language barriers in healthcare utilization provides a deeper understanding of the research findings.

When controlling for personal characteristics in the multinomial logistic model, the isolated effect of having spatial proximity to a community health center has virtually no impact on the likelihood that an individual claims a community health center as their regular source of care versus stating having no regular source of care, except for those with chronic medical conditions. This is an important distinction as a critical function of the medical safety net is to identify and support vulnerable populations before illness appears or control it before it worsens. The study findings support the latter, but the former requires pulling non- and sporadic users into the primary healthcare system, which is a shortcoming of our current system. This finding is also supported by the bivariate analysis, which suggested that there is no difference in the proportion of people who report having no regular source of healthcare for those with spatial proximity to a community health center and those without it. In this case, mere access to quality and affordable healthcare is not enough to promote regular use; vulnerable populations are not fully embracing this socially expected model of behavior, which in turn further worsens health disparities between at-risk populations and their more advantaged counterparts, leaving individuals even more marginalized without a regular source of care for addressing their health needs.

Conclusion

An essential component of the community health center program's ability to remain relevant and funded across presidential administrations has been the ability to point to a significant body of literature touting the success of the community health center model. Access to a community health center has been shown to be beneficial for users compared to non-users across various health measures, and barrier-free access to culturally competent, quality health services is the key to reducing health disparities between social groups. However, while efficacy studies tout the benefits of the community health center program and cite its record of disparity reductions between historically marginalized groups and whites, overall health disparities between social groups in the United States have persisted and, in some cases, worsened over the past five decades. The persistent stratification of health in the United States suggests that addressing the fundamental causes of health disparities is far more complex than solely relying on access alone. This chapter delves into this paradox, exploring the nuanced relationship between spatial proximity to a community health center and the likelihood of reporting regular use of it or other specific sources of care versus having no source of care at all. This inquiry is a necessary deviation from the traditional methods of measuring community health center efficacy that obscure a weak association between having access to a center and better health for the average person. Changing how we measure the impact of community health centers coupled with an administrative return to its poverty eradication roots could be an essential step in addressing health disparities for historically marginalized populations.

Overall, placing a community health center in a neighborhood does not automatically transform the health of those living there. It is critical to identify and understand which groups benefit the most from access and which do not so that specialized efforts can be made to focus more on pulling marginalized populations into care. This tactic of shifting from a safety net to a casting net could help to raise the overall health of communities, not just the tiny subset of the population who are compliant with social expectations around health and taking care of oneself. Adjusting to the needs of the community, pulling patients into care, getting the word out about the availability of services, and developing a concrete plan to combat medical distrust would go a long way in starting to combat the issue of health disparities for historically excluded and at-risk communities as compared with those having a relative social advantage. This emphasizes the importance of understanding the community's needs in healthcare planning.

Some limitations arise from this analysis and argument. One of the fundamental assumptions is that individuals tend to use healthcare services close to where they live. In reality, it is reasonable to assume that people not only travel considerable distances to seek the healthcare provider of their choice, but they may also find it convenient to use health services far from home but close to their school or workplace. In this analysis, spatial proximity is restricted to the respondent's residential zip code, which is limiting. Another limitation is that the utilization of community health centers or other healthcare services is not tracked in this analysis. While community health centers are not a regular source of care for individuals, they may use community health center services sporadically based on situational circumstances and the availability of other sources of healthcare. Finally, because the analysis is based on cross-sectional survey data, it is impossible to make a causal argument about the effect of spatial proximity to a community health center on health. Only the observed association between variables can be asserted. While these findings support current literature suggesting that spatial proximity to care is insufficient to transform at-risk populations into regular primary care users, it is essential to note that individuals may prefer to access primary care services outside of their immediate neighborhoods, potentially mediating the observed effect of proximity to care on the likelihood of having a regular source of care. Also, because this analysis is based on cross-sectional survey data, it is impossible to make a causal argument about the relationship between variables. Only the observed association can be assumed and used to inform future studies with little evidence for causation. Due to the complexity and prevalence of barriers to healthcare for vulnerable and at-risk populations, these findings suggest that improving geographic access to primary healthcare does not guarantee positive outcomes for target groups. The magnitude of social disadvantage on vulnerable and at-risk populations can have a devastating effect on health outcomes that is not easily overcome by social programs.

By primarily focusing on the risk factor of access, the concern for the more fundamental social factors influencing community health has become overshadowed by the need to support the nation's medical system. Access alone cannot transform individual and community health, especially in communities with concentrated personal and neighborhood disadvantages. Ideals of community development and social justice are inherent in the establishment of community health centers. Providing quality, low-cost healthcare is not just about health per see, but also about improving a neighborhood's social and economic structures. Overcoming this hurdle requires a multipronged approach to strengthening and expanding the community health center safety net while leveraging networks of like-minded community-based organizations already in place to pull more patients into care and systematically focus on community building. As the safety net for the larger medical system, community health centers are critical in ensuring that no one in their community falls through the cracks. As such, administrators must have the flexibility to widen their view beyond their compliant patients and take a closer look at their immediate neighbors who may not be using services because they think the center is not "for them" or that the centers are only good enough for acute emergencies, but not regular preventative care. The goal of the community health center program must be developing and implementing structural solutions to resolve health inequalities, not merely serving as a cost-saving measure for the broader healthcare system.

Conclusion

After the Last Dance: The Future of the Community Health Center Program's Image

Successful techniques in administrative leadership can be paradoxical when applied to strategies for the survivability of publicly supported agencies that depend on legislative authorizations. The community health center program offers a salient case study, presenting a story of success as well as caution for several reasons. First, having originally been conceived as a War on Poverty demonstration effort, the community health center program has not only survived numerous presidential and legislative challenges to its existence but has also thrived, showcasing its resilience and adaptability. As illustrated by the ebb and flow of health center support over various administrations and its multibillion-dollar budget today, community health centers have successfully adapted to the sociopolitical forces that have threatened their existence across several presidential administrations. The media has been a critical partner in their success in remaining relevant, as it has raised public awareness about the human experience of poverty and inequality and the benefits of the community health center program, especially when challenged nationally. The strong ties between the *New York Times* and community health center administrators and advocates over the past sixty years have helped keep community health centers on the public agenda within a policy subsystem that tends to overlook the healthcare needs of marginalized populations. The coverage built and supported by the *New York Times* in concert with administrators and advocates humanized the issue of health disparities in society.

More importantly, the coverage has highlighted the community health center program as a critical healthcare safety-net partner, helping to solidify the program's survival. Because of the administrative work undertaken to shape the program's image, it has been relatively easy to defend; it offers demonstrated benefits to communities in crisis and cost savings to taxpayers. The media has been indispensable in keeping the national public aware of these benefits, especially during times of policy change. Although the *New York Times* does not encompass all US media coverage of health programs for historically marginalized populations, it illustrates how socially responsible coverage can remind the public of its moral, economic, and social responsibilities.

Second, a primary strategy for the program's survival has been the deliberate management of its social image by agency administrators and advocates. This strategic approach ensures that the program remains palatable to elected officials and the public while keeping its primary mission intact. In the conclusion of her book about the history of the community health center program, Lefkowitz (2007) argues that "we are constantly admonished to learn from history, but as I have traveled around the country, I've also seen a jarring disconnect between the rich, emotional legacy of health centers and their need to compete in a modern marketplace as sophisticated providers" (p. 135). For decades, nonprofit organizations have been subject to environmental pressures for being more like for-profit companies. Funders and the public expect public sector agencies to be efficient, effective, and transparent as they meet their organizational objectives. To stay relevant and funded, administrators leading these agencies must be savvy and informed across multiple sectors, including nonprofit, government, and for-profit (Salamon, 2015). In the case of community health centers, upwards of 60 percent of their operational funds come from federal grants. As a result, they are highly dependent on the positive image of their work and the strength of advocacy efforts by individual legislators, lobbyists, and advocacy groups to ensure their relevancy and survival.

Third, although this administrative strategy successfully maintained and increased bipartisan funding support, community-level health disparities, a primary target of elimination for the program, persist in the areas it serves. It is important not to underestimate the positive outcomes of the community health center program. Its benefits are real, even if not as impactful for some as for others. Regular community health center patients have better access to quality healthcare and better health outcomes than nonpatients of similar demographics. According to federal reporting, patients of community health centers undergo more screenings, have better management of chronic conditions, and experience lower mortality rates than their non-patient peers, largely due to the centers' comprehensive and culturally competent care. Additionally, the community health center program has significantly reduced the number of uninsured individuals in its communities, leading to improved health outcomes and reduced healthcare costs. However, the measured successes of the program tend to focus on regular users as the unit of analysis, rather than the entire community that the community health center seeks to serve. Federal reporting requirements compel the centers to provide annual progress data for all patients within the centers' care; however, nonregular users and some noncompliant patients (such as people with diabetes who come for services less than twice a year) are generally excluded from the scope of the reporting (Bureau of Primary Health Care, 2022).

Community health centers are positioned as easily accessible, quality healthcare options available to anyone who seeks them out. When measuring and reporting programmatic impact, this centering of users makes sense—how can a social program expect to change outcomes for individuals who do not use it? However, a "built it and they will come" approach to healthcare overemphasizes the power of access and activates the sociological phenomenon of the sick role. The concept of the sick role is built on the assumption that those living with symptoms or awareness of being in a disease state will act in accordance with how society deems an ill person should act. Part of the expectation is that the person will seek care and eagerly take the opportunity to become a compliant patient if presented (Parsons, 1951). This commonly applied model overlooks the various social barriers that help to explain why an individual might choose to be a nonpatient, a position associated with deviance in our contemporary society.

With community health centers serving only a fraction of the historically marginalized populations in the United States, a large proportion of the population that could benefit from community health center programming but are not users (Politzer et al., 2001). These non-users may become further marginalized and stigmatized due to their perceived noncompliance, as program regulations do not require or fund centers to pull potential patients into their care. This focus on traditionally compliant patients, coupled with the emphasis on risk factors as opposed to the fundamental causes of disease, systematically leaves the most marginalized populations out of the safety-net process to the detriment of population health outcomes.

While this is a challenging concept, it is also crucial to the program's vision of eliminating the causes and consequences of poverty, which was one of the foundational goals of the program. If taking a communal approach, increasing the penetration and use of the program among the most marginalized populations should be a prominent operational goal, measured and celebrated at the federal level. Instead, measures for compliant patient success and cost savings in the external environment are among the most regularly reported wins. Congress is most impressed by the hundreds of millions of dollars saved in emergency room visits and uncompensated care. These health outcomes become rooted in the shortsighted belief that the mere improvement of access can transform sick communities. In reality, access alone is insufficient to improve population health for the most marginalized communities. It is crucial to reiterate the importance of the program's impact on marginalized populations, as this is a significant and pressing issue that requires immediate attention.

Fourth, by becoming increasingly concerned with being perceived as beneficial to mainstream society rather than focusing on transforming marginalized communities, the program has inadvertently de-emphasized the urgency of addressing the root causes and consequences of poverty. The shift from focus on the community to focus on the nation is an example of macro-level prioritization. It has led to a broader focus on supporting the overall healthcare system, potentially at the expense of addressing community-level health disparities. While the community health center mission demonstrates considerable sensitivity to the nonmedical risk factors that influence health, it falls short of engaging with the structural factors exacerbating poor health for vulnerable populations. This limited scope is concerning, especially considering the persistence of health disparities among socially marginalized groups. By primarily focusing on serving as a safety net for the benefit of the broader healthcare system, the program's initial focus on eradicating fundamental social causes of poor health becomes secondary and generally unaddressed.

As community health centers choose how elected officials and the public perceive their usefulness, the image of them catching and treating disease symptoms is prioritzed over reducing and eliminating the foundational social factors causing disease. Putting diagnostic testing and screenings over the amelioration of the fundamental social causes of disease can offer a superficial and temporary solution to the problem of poor health among historically marginalized communities. The implication of this shift is evident in the persistence of health disparities despite the billions of dollars invested annually into the healthcare safety net.

Finally, despite its undeniable success, a shift in focus toward mainstream society has inadvertently led to a de-emphasis on the root causes of poverty. Returning to the program's original community-based poverty eradication focus is not just important; it is crucial to effectively address community-level health disparities, a mission that should never be forgotten.

The community health center program's contemporary image as a safety net serving to fortify the greater healthcare system is a stark contrast to the program's early days. When the first centers opened in Mound Bayou and Roxbury, the focus was on transforming communities and helping the people who lived in those communities flourish and overcome poverty's challenges. Advocates of the community health center model were dedicated to innovative ways of addressing the social issues that affect community health, which was also directly linked with other anti-poverty initiatives at all levels of government. While risk factors for disease are easier to manage and report, which is crucial for the program's viability, the original mission of transforming the health of communities to eradicate poverty has been obscured.

Contemporarily positioned as a safety-net medical provider, it is crucial that the program return to its roots in neighborhood-based poverty-eradication in order to fundamentally address the social factors causing health disparities in the United States. Naturally, some centers nationwide have already committed to this ideal, never losing sight of this aspect of their work since their inception. These are not just places for healthcare; they are beacons of hope and extraordinary exemplars in the field. Without the proper funding and support from the federal government to support this extension of the work, it can only be done at the expense of agency administrators leading their organizations above and beyond their call of duty.

Where We Go from Here

It is a well-known adage in public administration that what gets measured and assessed gets prioritized in organizations. As with any federal program, reporting and standardization have become ubiquitous in the administration and management of community health centers. Administrators have to do more with less as they seek to meet the equity needs of their communities beyond what is required by federal guidelines. The financial challenges facing the community health center program are a cause for concern and require immediate attention to ensure its long-term viability and effectiveness. Staying in favor of Congress and the current Presidential administration is a good starting place for any federally subsidized program. Gradually swaying public opinion in the direction of social equity also goes a long way when educating about the role of community health centers in transforming outcomes for marginalized communities.

Public administrators are among the most creative, innovative, and committed organizational leaders across any of the three major economic sectors. Balancing the public's needs in highly transparent environments while also corralling elected officials and implementing public policy fairly and efficiently is no easy task. Public administrators must also be committed professionally to social justice through public service. Community health center administrators are at the forefront of this effort as public health takes center stage in the policy debate. However, inequality is progressively worsening between the bottom and top of the social stratification ladder, affecting health outcomes for all.

For community health center administrators to surpass their current achievements, legislative and public support for social equity advancement must be in place and backed by substantial financial resources. The current situation, where administrators are expected to do more with less, prioritizes necessary diagnostic activities over those that could enhance health equity. Society's commitment must extend beyond normative ideals to address health disparities. The next steps are relatively straightforward once the proper resources are in place. Community health center administrators and advocates should then be free to Heed the Data, Dialogue and Collaborate, and Act with Courage to achieve health equity goals in historically marginalized neighborhoods and communities once and for all.

To heed the data means taking meaningful action on the trends that data collection efforts reveal. Health researchers have long reported on the scope and nature of health inequity and disparities in the United States. Significant public funds must be invested in addressing the root causes of these problems, rather than primarily investing in the curative aspects of health.

Identifying the right interventions for the root cause of health disparities necessitates dialogue and collaboration across multiple fronts. Interagency collaboration at the federal level is crucial, as the issue of health disparities extends beyond healthcare and activates many public functions. Dialogue across various levels of government and intersectoral collaboration is also essential. Most importantly, the community must be at the heart of the indigent healthcare effort. Representatives should be active participants in developing healthcare solutions, going beyond occasional representation on advisory boards. Dialogue should not be limited to information gathering and active listening but should foster genuine interaction where each party deeply and respectfully listens to one another, leaving the interaction altered for the better. Healthcare administrators and advocates cannot assume what communities need; they must ask what is required and believe what they learn.

Finally, administrators must have the courage to take action. Advocating for and prioritizing marginalized communities should not be a career or funding killer. A culture of allyship and courage must be fostered, which can only happen as society becomes less polarized and more respectful of alternative opinions and perspectives. This is a tall order, but everyone has a responsibility to progress. These three steps are the future of the dance: simple and effortless for the observer but complex, sacrificial, and selfless for the performer.

Methodological Appendix

Chapter 3: Content Analysis Background and Procedures

For this media-based content analysis, print articles were identified from the *New York Times*: 1851– (ProQuest Historical Newspapers) database and the current *New York Times* website using the subject search function in temporal alignment with presidential administrations implementing major policy changes influencing indigent health policy. The *New York Times* is an ideal source for a media-based content analysis because of its leading roles both in covering national policy and setting the news agenda for other US media. Founded in 1851, the paper developed a strong tradition of investigative journalism, offering in-depth analysis of various social, economic, and political issues. The *New York Times* also reaches a broad, diverse audience both nationally and internationally, making it a valuable resource for studying public discourse on significant topics. Its wide readership includes influential policymakers, academics, and business leaders, all of whom contribute to shaping public opinion. By analyzing its content, researchers can gain insights into how media narratives are shaped, reinforced, and consumed by different segments of society. Moreover, *New York Times* editors maintain high editorial independence and quality control.

For this study, the article search was restricted to the term "community health center" and the selected years for each presidential term associated with community health center reauthorization dates. The findings were then pared down by reading the headline and lede to include only news reports and feature articles. A close reading of the remaining articles was performed, and all phrases and references to federal legislation were documented and tracked utilizing Atlas.ti version 24. A systemic trend analysis using the Atlas.ti software was then completed to uncover a theme across all articles from each Presidential administration.

Despite occasional criticisms of bias, the *New York Times* is generally regarded for striving to present multiple viewpoints, particularly in its opinion sections, which include perspectives from across the political spectrum. Its robust digital presence and archival resources make accessing historical data for longitudinal studies easier.

For the legislative and executive branch content analyses, the full text of the first State of the Union for each President serving from November 1963 to March 2024 was downloaded from the American Presidency Project website (https://www.presidency.ucsb.edu/) and uploaded to Atlas.ti version 24. A close reading of the speeches coupled with the administration overview information provided on the American Presidency Project website provided the thematic data needed for the analysis.

The Library of Congress was used as the source of the legislative content analysis using the Congress.gov website. Using the advanced search function, the term "community health center" in all fields, including bill text, was applied to all congressional sessions for the years 1951–2024. The full text of every result that became law and mentions the establishment, authorization, expansion, or funding of community health centers was downloaded from the website and uploaded to Atlas.ti version 24. The prompt "summarize the bill in three sentences" was then used within the smart coding function of Atlas.ti to provide summary text for the bill, which was then used for a close reading and subjective thematic analysis. The bill's full text was then revisited to obtain direct quotes and verify the automated coding schemes. The NACHC website, notably the "About Us" and "Community Health Center Milestones" tab, was used to verify that all relevant legislation was captured in the search.

Chapter 4: Quantitative Variables

One of the preformatted reports available in the HRSA data warehouse is a list of all HRSA Health Care Delivery Sites (HRSA, 2024). Filtering for the Philadelphia metropolitan area centers, a list of eighty-three active community health center service sites was generated in 2022. Using a combination of the available fields on HRSA export and web searches, the eighty-three sites were narrowed to forty-five that were in scope for this analysis. Inclusion criteria included having no restriction on the type of patient accepted for care, providing primary healthcare services, and being open and active during the time of the SPHHS survey collection. Then, zip codes in the community health center file were marked in the SPHHS file using the community health center Access variable coded "Y," indicating that respondents living in those zip codes have spatial proximity to a community health center, and "N," indicating no proximal community health center access. The resulting variable was a zip code-level dichotomous indicator that was then mapped to each SPHHS respondent.

In addition to the primary predictor of spatial proximity to a community health center, this analysis assessed the main and interactive effects of several secondary predictors on the likelihood of having a usual source of care, such as race, ethnicity, poverty status, community urbanicity, presence of a chronic medical condition, and insurance status. Race was captured in the SPHHS using six distinct categories: white, Black or African American, Asian or Pacific Islander, American Indian or Alaska Native, Biracial or Multiracial, or Something Else (specified by the respondent). The survey also recorded the respondent's ethnicity. To permit the comparison of Hispanics to other racial groups in this analysis, a race/ethnicity variable was created, which collapsed the data into four categories: white (not Hispanic), Black (not Hispanic), Hispanic, and Other (not Hispanic) in this study. Poverty Status was an imputed variable calculated from the household size and income variables in the SPHHS and compared with the federal poverty guidelines. Because community health centers provide subsidized services to patients with household incomes at or below 200 percent of poverty, the poverty threshold set in the analysis is 200 percent of poverty or below. Respondents residing in Philadelphia County were coded as living in an urban area, and respondents residing in the four remaining counties surrounding Philadelphia were coded as suburban residents. Chronic disease condition indicators were collected in the SPHHS and consisted of four chronic health conditions: Asthma, Diabetes, Arthritis, or Hypertension. A positive indication of at least one of these conditions translated into a positive indication for the chronic disease condition variable. Finally, insurance status was captured in the SPHHS and reflected whether the respondent had any insurance. The reminding predictors, age of the respondent, gender, and education were all captured in the SPHHS and were included as control variables in this analysis.

References

Andrulis, D. P. (1998). Access to Care Is the Centerpiece in the Elimination of Socioeconomic Disparities in Health. *Annals of Internal Medicine*, *129*(5), 412–16. https://doi.org/10.7326/0003-4819-129-5-199809010-00012.

Ayanian John Z., Kohler Betsy A., Abe Toshi, & Epstein Arnold M. (1993). The Relation between Health Insurance Coverage and Clinical Outcomes among Women with Breast Cancer. *New England Journal of Medicine*, *329*(5), 326–331. https://doi.org/10.1056/NEJM199307293290507

Bailey, M. J., & Danziger, S. (2013). *Legacies of the War on Poverty*. Russell Sage Foundation.

Bailey, M. J., & Goodman-Bacon, A. (2015). The War on Poverty's Experiment in Public Medicine: Community Health Centers and the Mortality of Older Americans. *The American Economic Review*, *105*(3), 1067.

Berkowitz, D., & TerKeurst, J. V. (1999). Community as Interpretive Community: Rethinking the Journalist-source Relationship. *Journal of Communication*, *49*(3), 125–36. https://doi.org/10.1111/j.1460-2466.1999.tb02808.x.

Berry, J. M. (2010). *The New Liberalism: The Rising Power of Citizen Groups*. Rowman & Littlefield.

Biden, J. (2022). *Address Before a Joint Session of the Congress on the State of the Union | The American Presidency Project*. https://www.presidency.ucsb.edu/documents/address-before-joint-session-the-congress-the-state-the-union-28.

Bird, D. (1967, April 4). Automated System Records Basic Patient Data: Facility at Lenox Hill Notes Temperature, Respiration and Blood Pressure. *The New York Times*, p. 36.

Bosman, J., Tavernise, S., & Baker, M. (2020, April 23). Why These Protesters Aren't Staying Home for Coronavirus Orders. *The New York Times*. https://www.nytimes.com/2020/04/23/us/coronavirus-protesters.html.

Boyle, A. (1996, December 1). Adding Reading to the Health Care Plan. *The New York Times*, CN22.

Broadwater, L. (2021, April 27). In a Locked-Down Capitol, Biden Will Address a Sparse Audience. *The New York Times*. https://www.nytimes.com/2021/04/27/us/politics/biden-joint-address-congress.html.

Brown, L. (1975, July 1). Clinic Is Losing Cable-TV Link to Doctors. *The New York Times*, p. 61.

Byrne, M., Murphy, A. W., Plunkett, P. K., McGee, H. M., Murray, A., & Bury, G. (2003). Frequent Attenders to an Emergency Department: A Study of Primary Health Care Use, Medical Profile, and Psychosocial Characteristics. *Annals of Emergency Medicine, 41*(3), 309–18. https://doi.org/10.1067/mem.2003.68.

Bryson, J. M. (2018). *Strategic Planning for Public and Nonprofit Organizations: A Guide to Strengthening and Sustaining Organizational Achievement*. John Wiley & Sons.

Bureau of Primary Health Care. (2022). *Uniform Data System Reporting Requirements for 2022 Health Center Data*. HRSA Health Center Program. https://bphc.hrsa.gov/sites/default/files/bphc/data-reporting/2022-uds-manual.pdf.

Bush, G. W. (1990). *Address Before a Joint Session of the Congress on the State of the Union | The American Presidency Project*. https://www.presidency.ucsb.edu/documents/address-before-joint-session-the-congress-the-state-the-union-22.

Bush, G. H. W. (2002). *Address Before a Joint Session of the Congress on the State of the Union | The American Presidency Project*. https://www.presidency.ucsb.edu/documents/address-before-joint-session-the-congress-the-state-the-union-2.

Carrillo, J. E., Carrillo, V. A., Perez, H. R., Salas-Lopez, D., Natale-Pereira, A., & Byron, A. T. (2011). Defining and Targeting Health Care Access Barriers. *Journal of Health Care for the Poor and Underserved, 22*(2), 562–75. https://doi.org/10.1353/hpu.2011.0037.

Carter, J. (1978). *The State of the Union Address Delivered Before a Joint Session of the Congress. | The American Presidency Project*. https://www.presidency.ucsb.edu/documents/the-state-the-union-address-delivered-before-joint-session-the-congress-1.

Casanova, C., & Starfield, B. (1995). Hospitalizations of Children and Access to Primary Care: A Cross-National Comparison. *International Journal of Health Services, 25*(2), 283–94. https://doi.org/10.2190/PCF7-ALX9-6CN3-7X9G.

Chen, B. K., Hibbert, J., Cheng, X., & Bennett, K. (2015). Travel Distance and Sociodemographic Correlates of Potentially Avoidable Emergency Department Visits in California, 2006–2010: An Observational Study. *International Journal for Equity in Health, 14*(1), 30. https://doi.org/10.1186/s12939-015-0158-y.

Clarke, J. N., McLellan, L., & Hoffman-Goetz, L. (2006). The Portrayal of HIV/AIDS in Two Popular African American Magazines. *Journal of Health Communication, 11*(5), 495–507. https://doi.org/10.1080/10810730600752001.

Clinton, W. (1993). *Address Before a Joint Session of the Congress on the State of the Union | The American Presidency Project*. https://www.presidency.ucsb.edu/documents/address-before-joint-session-the-congress-the-state-the-union-12.

Cohen, M. D., March, J. G., & Olsen, J. P. (1972). A Garbage Can Model of Organizational Choice. *Administrative Science Quarterly, 17*(1), 1–25. https://doi.org/10.2307/2392088.

Confessore, N. (2005, February 12). Among Gays Grown Complacent And Often Careless, Fear Returns. *The New York Times*, B4.

Cropf, R. A., & Wagner, J. L. (2018). *American Public Administration: Public Service for the Twenty-first Century* (Second Edition). Routledge.

Dairz, B. (2002, December 3). A Doctor's Journal: In Pain Clinic, Fruit, Candy and Relief. *The New York Times*, F1.

Decker, S. L. (2012). In 2011 Nearly One-Third of Physicians Said They Would Not Accept New Medicaid Patients, But Rising Fees May Help. *Health Affairs, 31*(8), 1673–9. https://doi.org/10.1377/hlthaff.2012.0294.

DeVoe, J. E., Fryer, G. E., Phillips, R., & Green, L. (2003). Receipt of Preventive Care Among Adults: Insurance Status and Usual Source of Care. *American Journal of Public Health, 93*(5), 786–91. https://doi.org/10.2105/AJPH.93.5.786.

Dor, A., Pylypchuck, Y., Shin, P., & Rosenbaum, S. (2008). Uninsured and Medicaid Patients' Access to Preventive Care: Comparison of Health Centers and Other Primary Care Providers. *Geiger Gibson/RCHN Community Health Foundation Research Collaborative*. https://hsrc.himmelfarb.gwu.edu/sphhs_policy_ggrchn/15.

Duffy, J. (1992). *The Sanitarians: A History of American Public Health*. University of Illinois Press.

Einstein Medical College Gets Million for Center. (1968, May 26). *The New York Times*, p. 72.

Farberman, R., McKillop, M., Lieberman, D. A., Delgado, D., Thomas, C., Cunningham, J., & McIntyre, Kevin. (2020). *The Impact of Chronic Underfunding on America's Public Health System: Trends, Risks, and Recommendations, 2020* [Issue Report]. Trust for America's Health. https://www.tfah.org/wp-content/uploads/2020/04/TFAH2020PublicHealthFunding.pdf.

Farmer, P. (2010). Partner to the Poor: A Paul Farmer Reader. In *Partner to the Poor*. University of California Press. https://doi.org/10.1525/9780520945630.

Finer, H. (1941). Administrative Responsibility in Democratic Government. *Public Administration Review, 1*(4), 335–50. https://doi.org/10.2307/972907.

Fisher, I. (1998, June 21). Health Care for Lesbians Gets a Sharper Focus. *The New York Times*, WH27.

Ford, G. (1975). *Address Before a Joint Session of the Congress Reporting on the State of the Union. | The American Presidency Project*. https://www.presidency.ucsb.edu/documents/address-before-joint-session-the-congress-reporting-the-state-the-union-1.

Friedrich, C. J. (1940). Public Policy and the Nature of Administrative Responsibility. *Public Policy, 1*, 1–20.

Fuentes, A. (2001, March). Clinic Casts Lifeline to Isolated Migrant Workers. *New York Times*, F5.

Gadson, D. N. (2023). Gaps in the Safety-Net: Community Health Centers' Diminutive Effect on Having a Regular Source of Care. In J. J. Kronenfeld (Ed.), *Social Factors, Health Care Inequities and Vaccination* (Research in the Sociology of Health Care, Vol. 40, pp. 123–36). Emerald Publishing Limited, Bingley.

Gadson, D. N., & Park, S. (2024). Social Equity and the Responsible Administrator: The Challenge of Equity in Public Policy Initiatives. *Administration & Society*, 00953997241249174. https://doi.org/10.1177/00953997241249174.

Galbraith, J. K. (1963). *The Affluent Society*. Penguin Books.

Gillette, M. L. (2010). *Launching the War on Poverty: An Oral History*. Oxford University Press.

Goldstein, J. (2021, January 8). Cuomo Widens Eligibility After Vaccine Goes Unused or Is Even Thrown Out. *The New York Times*. https://www.nytimes.com/2021/01/08/nyregion/nyc-coronavirus-vaccine-delays.html.

Graham, R., & Cave, D. (2023, August 25). After the Fires, Native Hawaiians Seek Revival Through Ritual. *The New York Times*. https://www.nytimes.com/2023/08/25/us/maui-hawaii-spiritual-healing.html.

Habermas, J. (1984). *The Theory of Communicative Action*. Beacon Press.

Hadley, J., Cunningham, P., & Hargraves, J. L. (2006). Would Safety-Net Expansions Offset Reduced Access Resulting From Lost Insurance Coverage? Race/Ethnicity Differences. *Health Affairs, 25*(6), 1679–87. https://doi.org/10.1377/hlthaff.25.6.1679.

Hansen, S. (2012, July). Hope in the Wreckage. *New York Times*, SM22.

Harrington, M. (1962). *The Other America*. Macmillan Publishing Company.

Health Resources & Services Administration. (2024). *Health Care and Service Delivery Sites*. https://data.hrsa.gov/topics/ESR.

Health-Clinic Strikers Are Resolute: Clinic Serves 85,000 "We're Willing to Negotiate." (1978, April 13). *The New York Times*, B2.

Hershenson, R. (1991, May 12). A Crisis at Prenatal Clinics in Yonkers: Clinics in Yonkers Face Crisis in Prenatal Care. *The New York Times*, WC1.

Hicks, L. S., O'Malley, A. J., Lieu, T. A., Keegan, T., Cook, N. L., McNeil, B. J., Landon, B. E., & Guadagnoli, E. (2006). The Quality Of Chronic Disease Care In U.S. Community Health Centers. *Health Affairs, 25*(6), 1712–23. https://doi.org/10.1377/hlthaff.25.6.1712.

Hodgetts, D., Chamberlain, K., Scammell, M., Karapu, R., & Waimarie Nikora, L. (2008). Constructing Health News: Possibilities for a Civic-oriented Journalism. *Health, 12*(1), 43–66. https://doi.org/10.1177/1363459307083697.

Hoffman, J. (2017, June 26). Sick and Afraid, Some Immigrants Forgo Medical Care. *The New York Times*. https://www.nytimes.com/2017/06/26/health/undocumented-immigrants-health-care.html.

Holgash, K., & Heberlein, M. (2019). Physician Acceptance Of New Medicaid Patients: What Matters And What Doesn't. *Health Affairs Forefront*. https://doi.org/10.1377/forefront.20190401.678690.

Hunter, M. (1964, May 16). Poverty War Debated: Johnson's Program Raises Doubts in Many Quarters and Nation's Conscience has Yet to be Aroused. *New York Times*, E5.

Hutchison, J., Thompson, M. E., Troyer, J., Elnitsky, C., Coffman, M. J., & Lori Thomas, M. (2018). The Effect of North Carolina Free Clinics on Hospitalizations for Ambulatory Care Sensitive Conditions among the Uninsured. *BMC Health Services Research, 18*(1), 280. https://doi.org/10.1186/s12913-018-3082-1.

Johnson, A. G. (2018). *Privilege, Power, and Difference* (Third edition). McGraw-Hill Education,.

Johnson, K., & Goodnough, A. (2020, April 5). Community Health Centers, Lifelines to the Poor, Are Forced to Cut Back. The New York Times, p. 17.

Johnson, L. (1964a). *Economic Report of The President* (330.973). United States Printing Office. https://www.presidency.ucsb.edu/sites/default/files/books/presidential-documents-archive-guidebook/the-economic-report-of-the-president-truman-1947-obama-2017/1964.pdf.

Johnson, L. (1964b). *Annual Message to the Congress on the State of the Union. | The American Presidency Project*. https://www.presidency.ucsb.edu/documents/annual-message-the-congress-the-state-the-union-25.

Kant, P., & Rushefsky, M. E. (2005). *The Politics of Public Health in the United States*. M.E. Sharpe.

Kark, S. L., & Steuart, G. W. (1962). *A Practice of Social Medicine. A South African Team's Experiences in Different African Communities*. E & S Livingstone Ltd.

Karlamangla, S. (2022, January 27). Despite Burnout, These California Health Care Workers Aren't Quitting. *The New York Times*. https://www.nytimes.com/2022/01/27/us/california-health-care-workers-burnout.html.

Kaysen, R. (2012, December 12). Health Centers Find Opportunity in Brownfields. *New York Times*, B8.

Kendi, I. X. (2020, April 14). Stop Blaming Black people for Dying of the Coronavirus. *The Atlantic*. https://www.theatlantic.com/ideas/archive/2020/04/race-and-blame/609946/.

Kim, E. J., Park, J. H., Yoon, S. J., Lee, S. J., Cha, S. I., Park, J. Y., Jung, T. H., & Kim, C. H. (2006). Relationship between Dyspnea and Disease Severity, Quality of Life, and Social Factor in Patients with Chronic Obstructive Pulmonary Disease. *Tuberculosis and Respiratory Diseases*, *60*(4), 397–403.

Kirby, J. B., & Sharma, R. (2017). The Availability of Community Health Center Services and Access to Medical Care. *Healthcare*, *5*(4), 174–82. https://doi.org/10.1016/j.hjdsi.2016.12.006.

Klinenberg, E. (2015). *Heat Wave: A Social Autopsy of Disaster in Chicago*. University of Chicago Press.

Lefkowitz, M. B. (2007). *Community Health Centers: A Movement and the People Who Made It Happen* (1st edition). Rutgers University Press.

Liaw, W., Jetty, A., Petterson, S., Bazemore, A., & Green, L. (2018). Trends in the Types of Usual Sources of Care: A Shift from People to Places or Nothing at All. *Health Services Research*, *53*(4), 2346–67. https://doi.org/10.1111/1475-6773.12753.

Lindblom, C. E. (1959). The Science of "Muddling Through." *Public Administration Review*, *19*(2), 79–88. https://doi.org/10.2307/973677.

Link, B. G., & Phelan, J. (1995). Social Conditions As Fundamental Causes of Disease. *Journal of Health and Social Behavior*, 80–94. https://doi.org/10.2307/2626958.

Liu, J. (Jason). (2007). Health Professional Shortage and Health Status and Health Care Access. *Journal of Health Care for the Poor and Underserved*, *18*(3), 590–8. https://doi.org/10.1353/hpu.2007.0062.

Long, N. E. (1949). Power and Administration. *Public Administration Review*, *9*(4), 257–64. https://doi.org/10.2307/972337.

Lukpat, A. (2021, October 15). Hate Crimes and Pandemic Lead More Asian Americans to Seek Therapy. *The New York Times*. https://www.nytimes.com/2021/10/15/us/asian-american-therapy-hate-crimes.html.

Lynch, J. W., Smith, G. D., Kaplan, G. A., & House, J. S. (2000). Income Inequality and Mortality: Importance to Health of Individual Income, Psychosocial Environment, or Material Conditions. *BMJ*, *320*(7243), 1200–4. https://doi.org/10.1136/bmj.320.7243.1200.

Macintyre, S., McKay, L., & Ellaway, A. (2006). Lay Concepts of the Relative Importance of Different Influences on Health; Are there Major Socio-demographic Variations? *Health Education Research*, *21*(5), 731–9. https://doi.org/10.1093/her/cyl015.

Malone, R. E., Boyd, E., & Bero, L. A. (2000). Journalists' Constructions of Passive Smoking as a Social Problem. *Social Studies of Science, 30*(5), 713–35.

Martínez, A. D. (2019). Overcoming Institutional Barriers Faced by Community-Based Healthcare Institutions. In S. L. Arxer & J. W. Murphy (Eds.), *Community-Based Health Interventions in an Institutional Context* (pp. 159–178). Springer International Publishing. https://doi.org/10.1007/978-3-030-24654-9_12.

Marx, F. M. (1949). Administrative Ethics and the Rule of Law. *American Political Science Review, 43*(6), 1119–44. https://doi.org/10.2307/1950510.

May, J. H., Cunningham, P. J., & Hadley, J. (2004). *Most Uninsured People Unaware of Health Care Safety Net Providers.* Center for Studying Health System Change. *90*, 4.

Mccombs, M. E., & Shaw, D. L. (1972). The Agenda-setting Function of Mass Media. *Public Opinion Quarterly, 36*(2), 176–87. https://doi.org/10.1086/267990.

McEwen, B. S. (2006). Protective and Damaging Effects of Stress Mediators: Central Role of the Brain. *Dialogues in Clinical Neuroscience, 8*(4), 367–81.

McMorrow, S., & Zuckerman, S. (2014). Expanding Federal Funding to Community Health Centers Slows Decline in Access for Low-Income Adults. *Health Services Research, 49*(3), 992–1010. https://doi.org/10.1111/1475-6773.12141.

Medical Center Opens in Bronx: 3,000 Poor Children Will Be Given Care at Facility. (1968, December 2). *The New York Times*, p. 26.

Melvin, T. (1981, November 15). Impact of Cutbacks on Health-Care Sites. *The New York Times,* WC13.

Merzel, C., & Moon-Howard, J. (2002). Access to Health Services in an Urban Community: Does Source of Care Make a Difference? *Journal of Urban Health, 79*(2), 186–99. https://doi.org/10.1093/jurban/79.2.186.

Miller, A., Cunningham, M., & Ali, N. (2013). Bending the Cost Curve and Improving Quality of Care in America's Poorest City. *Population Health Management, 16*(S1), S–17. https://doi.org/10.1089/pop.2013.0038.

Miller, H. (1963, April 21). New Definition of Our "Poor": They Exist in Large Numbers in the Midst of Today's Affluence, though the Criteria for Judging them Differ Radically from Yesterday. *New York Times (1923–Current File)*, SM6.

National Association of Community Health Centers. (2023). *Community Health Center Chartbook 2023.* https://www.nachc.org/wp-content/uploads/2023/07/Community-Health-Center-Chartbook-2023-2021UDS.pdf.

Netherland, J., & Hansen, H. B. (2016). The War on Drugs that Wasn't: Wasted Whiteness, "Dirty Doctors," and Race in Media Coverage of Prescription Opioid Misuse. *Culture, Medicine and Psychiatry, 40*(4), 664–86. https://doi.org/10.1007/s11013-016-9496-5.

Nguyen, T., & Hirota, S. (2018, September 26). Opinion | Trump's Next Target: Legal Immigrants. *The New York Times*. https://www.nytimes.com/2018/09/25/opinion/trumps-next-target-legal-immigrants.html.

Obama, B. (2009). *Address Before a Joint Session of the Congress on the State of the Union | The American Presidency Project*. https://www.presidency.ucsb.edu/documents/address-before-joint-session-the-congress-the-state-the-union-17.

Office of Policy Planning and Research United States Department of Labor. (1965, March). *The Negro Family: The Case for National Action*. DOL. https://www.dol.gov/general/aboutdol/history/webid-moynihan

O'Malley, A. S., Forrest, C. B., Politzer, R. M., Wulu, J. T., & Shi, L. (2005). Health Center Trends, 1994–2001: What Do They Portend for the Federal Growth Initiative? *Health Affairs*, *24*(2), 465–72.

Parsons, T. (1951). *The Social System*. Free Press.

Pearlin, L. I. (1989). The Sociological Study of Stress. *Journal of Health and Social Behavior*, *30*(3), 241–56. https://doi.org/10.2307/2136956.

Pears, R. (1991, April 3). Higher Fees for Doctors May Not Help Medicaid. *New York Times*, A18.

Politzer, R. M., Schempf, A. H., Starfield, B., & Shi, L. (2003). The Future Role of Health Centers in Improving National Health. *Journal of Public Health Policy*, *24*(3), 296–306. https://doi.org/10.2307/3343376.

Politzer, R. M., Yoon, J., Shi, L., Hughes, R. G., Regan, J., & Gaston, M. H. (2001). Inequality in America: The Contribution of Health Centers in Reducing and Eliminating Disparities in Access to Care. *Medical Care Research and Review*, *58*(2), 234–48. https://doi.org/10.1177/107755870105800205.

Public Health Management Corporation. (2024). *Community Health Database—PHMC Research & Evaluation Group*. https://research.phmc.org/products/community-health-database.

Raudenbush, D. (2020). *Health Care Off the Books*. University of California Press.

Reagan, R. (1982). *Address Before a Joint Session of the Congress Reporting on the State of the Union | The American Presidency Project*. https://www.presidency.ucsb.edu/documents/address-before-joint-session-the-congress-reporting-the-state-the-union-2.

Regan, J., Schempf, A. H., Yoon, J., & Politzer, R. M. (2003). The Role of Federally Funded Health Centers in Serving the Rural Population. *The Journal of Rural Health*, *19*(2), 117–124. https://doi.org/10.1111/j.1748-0361.2003.tb00552.x

Rosano, A., Loha, C. A., Falvo, R., van der Zee, J., Ricciardi, W., Guasticchi, G., & de Belvis, A. G. (2013). The Relationship between Avoidable Hospitalization and

Accessibility to Primary Care: A Systematic Review. *European Journal of Public Health, 23*(3), 356–60. https://doi.org/10.1093/eurpub/cks053.

Ross, C. E., Mirowsky, J., & Pribesh, S. (2001). Powerlessness and the Amplification of Threat: Neighborhood Disadvantage, Disorder, and Mistrust. *American Sociological Review, 66*(4), 568–91. https://doi.org/10.1177/000312240106600405.

Rubin, I. (2000). *The Politics of Public Budgeting: Getting and Spending, Borrowing and Balancing: Vol. 4th ed.* Chatham House Publishing. https://ezproxy.villanova.edu/login?URL=https://search.ebscohost.com/login.aspx?direct=true&db=nlebk&AN=35221&site=ehost-live&scope=site.

Rule, S. (1981, November 17). A Bronx Health Center Battling for U.S. Funds: Patients Join Battle Alternative to Emergency Care Another Grant Threatened Fees May Go to $75 a Visit Other Centers Lack Staff Doctor Is Distressed. *The New York Times,* B3.

Sack, K. (2009, September 9). A Summer of Work Opens Medical Students' Eyes. *New York Times,* A18.

Salamon, L. M. (2015). *The Resilient Sector Revisited: The New Challenge to Nonprofit America.* Brookings Institution Press.

Saloner, B., Wilk, A. S., & Levin, J. (2020). Community Health Centers and Access to Care Among Underserved Populations: A Synthesis Review. *Medical Care Research and Review, 77*(1), 3–18. https://doi.org/10.1177/1077558719848283.

Sardell, A. (1988). *The U.S. Experiment in Social Medicine: The Community Health Center Program, 1965–1986.* University of Pittsburgh Press.

Saver, B. G., & Peterfreund, N. (1993). Insurance, Income, and Access to Ambulatory Care in King County, Washington. *American Journal of Public Health, 83*(11), 1583–8.

Sheehy, G. (1979, December 27). Hers. *The New York Times,* C2.

Shi, L., Stevens, G. D., & Politzer, R. M. (2007). Access to Care for U.S. Health Center Patients and Patients Nationally: How Do the Most Vulnerable Populations Fare? *Medical Care, 45*(3), 206–13.

Shi, L., Stevens, G. D., Wulu, J. T., Politzer, R. M., & Xu, J. (2004). America's Health Centers: Reducing Racial and Ethnic Disparities in Perinatal Care and Birth Outcomes. *Health Services Research, 39*(6p1), 1881–902. https://doi.org/10.1111/j.1475-6773.2004.00323.x.

Shi, L., Tsai, J., Higgins, P. C., & Lebrun, L. A. (2009). Racial/Ethnic and Socioeconomic Disparities in Access to Care and Quality of Care for US Health Center Patients Compared With Non-Health Center Patients. *The Journal of Ambulatory Care Management, 32*(4), 342–50. https://doi.org/10.1097/JAC.0b013e3181ba6fd8.

Shin, P., Jones, K., & Rosenbaum, S. J. (2003). *Reducing Racial and Ethnic Health Disparities: Estimating the Impact of High Health Center Penetration in Low-income Communities*. Center for Health Services Research and Policy, The George Washington University.

Sierpina, D. (1995, November 26). Where Community is the Key on the Way to Health. *New York Times*, Section 13, p. 1.

Strelnick, A. H. (2004). Increasing Access to Health Care and Reducing Minority Health Disparities: A Brief History and the Impact of Community Health Centers Increasing Access to Health Care: Methods to Address the National Crisis. *New York University Journal of Legislation and Public Policy*, *8*(1), 63–80.

Sullivan, R. (1978, February 5). U.S. Sends Doctors To Inner-City Areas: Doctors Sent to Inner Cities. *The New York Times,* NJ1.

Tarraf, W., Vega, W., & González, H. M. (2014). Emergency Department Services Use Among Immigrant and Non-immigrant Groups in the United States. *Journal of Immigrant and Minority Health, 16*(4), 595–606. https://doi.org/10.1007/s10903-013-9802-z.

Taylor, T., & Salyakina, D. (2019). Health Care Access Barriers Bring Children to Emergency Rooms More Frequently: A Representative Survey. *Population Health Management*, *22*(3), 262–71. https://doi.org/10.1089/pop.2018.0089.

Tolchin, M. (1966, February 25). Brooklyn to Get a Day Hospital: Patients at 265-Bed Facility Actually to Live at Home, *The New York Times*, p. 13.

Trump, D. (2018). State of the Union Addresses | The American Presidency Project. Retrieved February 13, 2024, from https://www.presidency.ucsb.edu/documents/app-categories/spoken-addresses-and-remarks/presidential/state-the-union-addresses?field_docs_start_date_time_value%5Bvalue%5D%5Bdate%5D=2018.

Tuller, D. (2001, October 16). Experts Fear a Risky Recipe: Viagra, Drugs and H.I.V.: Fear of a Side Effect from the Drug's Casual Use: An Increase in Unsafe Sex. Fears of Risky Mix: Viagra, Drugs and H.I.V. *The New York Times,* F5.

Walker, M. (2021, February 18). Native Americans 'Left Out in the Cold' Under Trump Press Biden for Action. *The New York Times*. https://www.nytimes.com/2021/02/18/us/politics/native-americans-biden.html.

Weible, C. M., Heikkila, T., deLeon, P., & Sabatier, P. A. (2012). Understanding and Influencing the Policy Process. *Policy Sciences*, *45*(1), 1–21. https://doi.org/10.1007/s11077-011-9143-5.

Weiland, N. (2022, April 9). In This Michigan County, Pandemic Stimulus Funds Are Remaking Public Health Programs. *The New York Times*. https://www.nytimes.com/2022/04/09/us/politics/michigan-pandemic-stimulus.html.

Wen, M., Browning, C. R., & Cagney, K. A. (2003). Poverty, Affluence, and Income Inequality: Neighborhood Economic Structure and its Implications for Health. *Social Science & Medicine*, *57*(5), 843–60. https://doi.org/10.1016/S0277-9536(02)00457-4.

Wilson, W. (1887). The Study of Administration. *Political Science Quarterly*, *2*(2), 197–222. https://doi.org/10.2307/2139277.

Xu, K. T. (2002). Usual Source of Care in Preventive Service Use: A Regular Doctor versus a Regular Site. *Health Services Research*, *37*(6), 1509–29. https://doi.org/10.1111/1475-6773.10524.

Index

About the Author

Danielle N. Gadson, PhD, MPA, is a public administrator and medical sociologist with more than a decade of experience in community health center operations. A graduate of the Wharton School of Business at the University of Pennsylvania, Villanova University, and Temple University, Dr. Gadson is a full-time faculty member in the Department of Public Administration at Villanova University and a part-time lecturer and operations and equity consultant in the Philadelphia metropolitan area. Her research agenda focuses on the efficacy of public policy initiatives and structures, especially in service to historically marginalized populations. Through her work, Dr. Gadson highlights best practices for successfully balancing equity and efficiency in public policy administration.

www.ingramcontent.com/pod-product-compliance
Ingram Content Group UK Ltd.
Pitfield, Milton Keynes, MK11 3LW, UK
UKHW020208290926
15470UKWH00001B/64

9 781666 926699